The NICE Cyber Security Framework

Izzat Alsmadi · Chuck Easttom ·
Lo'ai Tawalbeh

The NICE Cyber Security Framework

Cyber Security Management

 Springer

Izzat Alsmadi
Department of Computing
and Cyber Security
Texas A&M University–San Antonio
San Antonio, TX, USA

Chuck Easttom
Plano, TX, USA

Lo'ai Tawalbeh
Department of Computing
and Cyber Security
Texas A&M University–San Antonio
San Antonio, TX, USA

ISBN 978-3-030-41989-9   ISBN 978-3-030-41987-5   (eBook)
https://doi.org/10.1007/978-3-030-41987-5

This Springer imprint is published by the registered company Springer Nature Switzerland AG
The registered company address is: Gewerbestrasse 11, 6330 Cham, Switzerland

Preface

In this book, we covered different subjects in cyber security based on the USA NIST NICE framework. We adopted their KSAs (Knowledge, Skills and Abilities). Our goal was to provide one possible instance of their template framework.

The book includes 12 chapters: Chap. 1: Information Assurance/ Encryption, Chap. 2: Information Systems Security Management, Chap. 3: IT Risk and Security Management, Chap. 4: Criminal Law, Chap. 5: Network Management, Chap. 6: Risk Management, Chap. 7: Software Management, Chap. 8: System Administration, Chap. 9: System Architecture, Chap. 10: Threat Analysis, Chap. 11: Training, Education, and Awareness, and Chap. 12: Vulnerability Assessment and Management.

As the field and the subjects in this area evolve rapidly, we tried to cover content that is current and includes enough details to help readers know enough details. KSAs in NICE framework are not symmetric where some of them maybe covered briefly and some others may require much more significant details and size.

With the intention to keep learning, we hope our readers will provide us with their feedback.

San Antonio, USA   Authors

Contents

Chapter 1
Information Assurance/Encryption

Chuck Easttom

General Cryptography Knowledge

A general working knowledge of cryptography is important to anyone working in cybersecurity. The NICE framework recognizes this. As you can see from the objectives for this chapter, there are several knowledge areas in the NICE framework that are devoted to cryptography. In this chapter you will learn about these areas. Algorithms and applications will be covered. A deep math background is not required, as the NICE framework does not require a mathematical understanding of cryptography, but merely the ability to apply cryptographic solutions. However, in this section we will cover sufficient cryptography to allow you to understand issues related to implementing cryptographic solutions.

Ancient Ciphers

Ancient ciphers are not directly addressed in the NICE framework. However, they are the standard place to begin teaching cryptography. These algorithms are not secure and cannot be used in a modern environment. However, they serve two objectives. The first is to provide an historical context for studying modern cryptography. The second is to allow the reader to become comfortable with the concepts of cryptography, before delving deeper into modern methods. Essentially, these algorithms are easier to understand, and thus provide a good starting place.

© The Editor(s) (if applicable) and The Author(s), under exclusive license to Springer Nature Switzerland AG 2020
I. Alsmadi et al., *The NICE Cyber Security Framework*,
https://doi.org/10.1007/978-3-030-41987-5_1

The Caesar Cipher

One of the oldest recorded ciphers is the Caesar cipher. This makes it a logical place to begin exploring cryptography. Many books on cryptography start with explaining the Caesar cipher, so we will follow that same approach here. The name of the cipher is based on a claim that this method was used by ancient Roman emperors, particularly Julius Caesar. This method is simple to implement, requiring no technological assistance. You choose some number by which to shift each letter of a text. For example, if the text is

> Cryptography
> And you choose to shift by two letters to the right then the message becomes
> Etarvqitcrja
> Or, if you choose to shift by five letter2 to the right, it becomes
> Hwduytlwfumd

It is claimed that Julius Caesar used 2 to the right. However, as with varying the number to shift by, you may also choose to shift to the right or left. If the shift leads to the end of the alphabet, then simply wrap around. So, the letter *Y* in a shift of two to the right would become *A*. You may use any shift you like. Regardless of the shift use, the Caesar cipher is part of a class of ciphers known as single-substitution ciphers, also called mono-alphabet ciphers. These names stem from the fact that these ciphers will always have a single substation for any given plaintext letter. So, in the case of shifting one to the left, *A* will always become *Z*. This makes them rather easy to break. But recall, this was first used at a time when most of the population was illiterate.

ROT 13

ROT 13 is another single alphabet substitution cipher. It is essentially a Caesar cipher that always shifts the same amount. All characters are rotated 13 characters through the alphabet.

The word Cryptography Becomes pelcgbtencul

ROT 13 is a single-substitution cipher.

This is actually used today in the Windows registry. As one example, Userassist key in the registry uses ROT-13 to obscure its contents. This may seem quite insecure, but it is not really meant for robust security. The registry is only accessible to someone who already has local administrative privileges. The real purpose is just to prevent a novice administrator from altering something he or she should not.

Atbash Cipher

In ancient times, this cipher was used by Hebrew scribes encrypting religious texts. Using it is simple; you just reverse the alphabet. This is a single substitution cipher, and by modern standards is quite easy to break. It simply reverses the alphabet; for example (in English), *A* becomes *Z*, *B* becomes *Y*, *C* becomes *X*, etc.

Multi-alphabet Substitution

Even in the era before computers, the flaws in single substitution ciphers became apparent. This led to poly-alphabet substitution. There have been many variations on this. But they all start from the same premise. If you have several different substitutions for a given plaintext, you disrupt the letter and word frequency making it harder to decrypt.

With Caesar you have a single shift, for example shift to the right 2 (+2). So, consider a simple example of poly alphabet substitution. Perhaps with three shifts: one to the right (+1), one to the left (−1), and two to the right (+2). As an example, we take the plaintext word: Chair

The first letter is shifted one to the right, yielding: Dhair
The second letter is shifted one to the left, yielding: Dgir
The third letter is shifted two to the right, yielding: Dgkr
Now since we have only three shifts, and four letters, we go back to the first shift, and the fourth letter is shifted one to the right, yielding: Dgks

If you have more letters, you continue with the three shifts. It should be fairly obvious that the more substitution ciphers you have, the more secure the cipher is. It should be noted, however, that this would still not be secure against modern computers.

One of the most widely known multi-alphabet ciphers was the Vigenère cipher. This cipher was invented in 1553 by Giovan Battista Bellaso. It is a method of encrypting alphabetic text by using a series of different mono-alphabet ciphers selected based on the letters of a keyword. This algorithm was later misattributed to Blaise de Vigenère, and so it is now known as the "Vigenère cipher," even though Vigenère did not really invent it.

Multi-alphabet ciphers are more secure than single-substitution ciphers. However, they are still not acceptable for modern cryptographic usage. Computer-based cryptanalysis systems can crack historical cryptographic methods (both single alphabet and multi-alphabet) easily. The single-substitution and multi-substitution alphabet ciphers are discussed just to show you the history of cryptography, and to help you get an understanding of how cryptography works.

Specific Modern Algorithms

Understanding cryptography requires some level of understanding of the particular algorithms used. It is not necessary that you become a cryptographer or mathematician, but it is important that you have a fundamental understanding of algorithms. In this section you will be presented with broad coverage of algorithms.

Symmetric Cryptography

Symmetric ciphers are those which use the same key to encrypt and to decrypt. They are always faster than asymmetric algorithms and provide security with much smaller keys. Symmetric ciphers are chosen for encrypting files, hard drives, and similar tasks. Symmetric ciphers can be further divided into two sub-groups. The first sub-group is block ciphers. The plaintext is divided into blocks of a fixed size, and each block is then encrypted. Stream ciphers encrypt the bits (or bytes) as they are input into the algorithm. In this section several algorithms will be examined. The more widely used an algorithm is, or its importance in the history of cryptography will determine how much detail is provided.

Key Schedule

Before we discuss specific algorithms, we should first discuss key schedule algorithms. All symmetric ciphers have a second algorithm that derives a different key for each round of the algorithm. These are called key scheduling algorithms. The concept is to take the symmetric key that the two parties have agreed upon, and to change it slightly each round of the algorithm. That leads to more change in the cipher text and greater security of the algorithm. The specifics of the key schedule algorithms are unique to each symmetric algorithm. DES uses a different key schedule than does AES. However, let us consider a simplified example to illustrate the concept.

Imagine a cipher that uses an 8-bit key (yes that would be absurdly small and easily brute forced, but this is just to illustrate a concept). Now you could simply use the same 8-bit key each round, but that would not be as robust as changing it slightly. So, you devise a simple key scheduling algorithm. Each round you key will be bit shifted to the right by 1. Thus, the initial symmetric key is: 11011000.

Round 1 it is first shifted to the right 1 and you are doing a shift with carry (i.e. the value at the end wraps around as opposed to just dropping off). Therefore, your first-round key is: 01101100.

The second-round key is: 00110110.

And this continues each round. Now this is a rather simplified illustration of the concept. The key schedule algorithms used in real ciphers are a bit more complex than this. However, this does illustrate the concept.

Data Encryption Standard

This is the oldest modern symmetric cipher. By modern, it is meant to differentiate from ancient ciphers like Caesar, Atbash, etc. DES uses a 56-bit key on a 64-bit block. Due to the key size, it is no longer considered secure. However, it is pivotal in the history of cryptography. Furthermore, its structure is used in many current symmetric algorithms.

DES was developed by Horst Feistel at IBM, working in conjunction with the U.S. Government. The algorithm is based on Horst Feistel's earlier algorithm named Lucifer. The role of the U.S. government was to help with the substitution boxes in DES. A substitution box (often called an s-box) is a hard-coded lookup table that substitutes some output value for the input bits. This is a part of all block ciphers. DES has eight different s-boxes, which will be discussed in this section. Figure 1.1 is one of those s-boxes.

The DES algorithm uses a structure called a Feistel Function (often called a Feistel network) that is now commonly used in many different block ciphers. The process is to take the input block (in the case of DES, that is a 64-bit block) and divide it into two halves. The right half will be input into some function, called a round function. It is called a round function because it is executed each round of the algorithm. This is shown in Fig. 1.2.

	x0000x	x0001x	x0010x	x0011x	x0100x	x0101x	x0110x	x0111x	x1000x	x1001x	x1010x	x1011x	x1100x	x1101x	x1110x	x1111x
0yyyy0	14	4	13	1	2	15	11	8	3	10	6	12	5	9	0	7
0yyyy1	0	15	7	4	14	2	13	1	10	6	12	11	9	5	3	8
1yyyy0	4	1	14	8	13	6	2	11	15	12	9	7	3	10	5	0
1yyyy1	15	12	8	2	4	9	1	7	5	11	3	14	10	0	6	13

Fig. 1.1 DES S-Box

Fig. 1.2 Feistel network

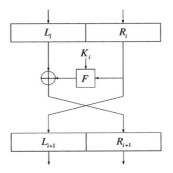

This structure is commonly called a Feistel Network. The number of rounds that are used and the details of the round function vary between algorithms. DES uses 16 rounds. The details of the round function are beyond the scope of this book. However, they are not overly complicated and you can find them in several different websites.

3DES

Eventually it became obvious that the key size for DES made it no longer secure. Thus, a search was on by the NIST to find a replacement. But that search would take time. In the interim, encrypting with DES three times was used as an alternative.

There were variations of 3DES that used only two keys. The text was first encrypted with key A. The cipher text from that operation was then encrypted with key B. Then the cipher text from that operation was encrypted, this time reusing key A. The reason for this is that creating good cryptographic keys is computationally intensive.

AES

This algorithm is one of the most widely used symmetric ciphers in the world today. The Rijndael block cipher was developed by two Belgian cryptographers, Joan Daemen and Vincent Rijmen. John Daeman. Rijndael was the algorithm eventually chosen to replace DES and become the Advanced Encryption Standard (AES). AES is a block cipher that works on 128-bit blocks. It can have one of three key sizes of 128, 192, or 256 bits. This was selected by the United States government to be the replacement for DES and is published as FIPS (Federal Information Processing Standard)-197.

The process to select the replacement for DES was a 5-year process involving 15 competing algorithms. Other algorithms that did not win that competition include such well-known algorithms as Twofish. The importance of AES cannot be overstated. It is widely used around the world and is perhaps the most widely used symmetric cipher. Of all the algorithms in this chapter, AES is the one you should give the most attention to.

AES can have three different key sizes: 128, 192, and 256 bits. The three different implementations of AES are referred to as AES 128, AES 192, and AES 256. The block size can also be 128, 192, or 256 bits. It should be noted that the original Rijndael cipher allowed for variable block and key sizes in 32-bit increments. However, the U.S. government uses these three key sizes with a 128-bit block as the standard for AES.

For those readers who want more detail, here is a general overview of the process used in AES.

The algorithm consists of a few relatively simple steps that are used during various rounds. The steps are described here:

- AddRoundKey—each byte of the state is combined with the round key using bitwise xor. This is where Rijndael applies the round key generated from the key schedule.
- SubBytes—a nonlinear substitution step where each byte is replaced with another according to a lookup table. This is where the contents of the matrix are put through the *s*-boxes. Each of the *s*-boxes is 8 bits.
- ShiftRows—a transposition step where each row of the state is shifted cyclically a certain number of steps. In this step the first row is left unchanged. Every byte in the second row is shifted one byte to the left (with the far left wrapping around). Every byte of the third row is shifted two to the left, and every byte of the fourth row is shifted three to the left (again with wrapping around.
- MixColumns—a mixing operation which operates on the columns of the state, combining the four bytes in each column. In the MixColumns step, each column of the state is multiplied with a fixed polynomial.

With the aforementioned steps in mind, this is how those steps are executed in the Rijndael cipher. For 128-bit keys, there are 10 rounds. For 192-bit keys there are 12 rounds. For 256-bit keys there are 14 rounds. Larger keys also have more rounds, which means they are strong based on both the larger key size and the additional rounds.

Now the Rijndael cipher has specific phases, the initial round, the rounds phase, and the final round. The aforementioned steps are applied in these stages, described below.

Initial Round

This initial round will only execute the addroundkey step. This is simply XOR'ing with the round key. This initial round is executed once, then the subsequent rounds will be executed. Thus, the algorithm begins with a basic exclusive or operation.

Rounds

This phase of the algorithm executes several steps, in the following order:

SubBytes
ShiftRows
MixColumns
AddRoundKey.

These steps were described in more detail previously.

Final Round

This round has everything the rounds phase has, except no mix columns.

- SubBytes
- ShiftRows
- AddRoundKey

As you can see, each individual step is not particularly complex. However, the culmination of all of these steps yields a very secure algorithm. We have not discussed the Rijndael *s*-box nor how it is derived. Nor have we discussed the key schedule algorithm. Those topics are beyond the scope of this current text. If you wish to delve deeper into AES, there is a wonderful animation that walks the viewer through every step, visually displaying the steps. It even describes the key schedule algorithm. You can find it in multiple locations on the internet (just search for AES animation), and it has been converted to a YouTube video https://www.youtube.com/watch?v=gP4PqVGudtg.

Blowfish

Blowfish is a symmetric block cipher published in 1993 and was intended as a replacement for DES. It was developed by a team lead by Bruce Schneier. It uses a variable-length key ranging from 32 to 448 bits. It is widely considered a robust and secure symmetric cipher. It is also distributed free of charge and without any patents or copyrights, thus making it attractive to budget-conscious organizations.

Serpent

Serpent was also a competitor for the AES competition and made it to the final 5, but was not selected. Serpent has a block size of 128 bits and can have a key size of 128, 192, or 256 bits, much like AES. The algorithm has a structure very much like AES. It uses 32 rounds regardless of the key size selected. Serpent was designed so that all operations can be executed in parallel. This is one reason it was not selected as a replacement for DES. At the time many computers had difficulty with the parallel processing. However, modern computers have no problem with parallel processing, so Serpent is once again an attractive choice.

RC4

RC4 is a stream cipher developed by Ron Rivest. The RC is an acronym for Ron's Cipher or sometimes Rivest's Cipher. There are other RC versions, such as RC5 and RC6. RC4 has been one of the most widely used stream ciphers. Stream ciphers are a bit less common that block ciphers but are still important to have some knowledge of.

FISH

This algorithm was published by the German engineering firm Seimans in 1993. The FISH (FIbonacci SHrinking) cipher is a stream cipher using Lagged Fibonacci generator along with a concept borrowed from the shrinking generator ciphers. A Lagged Fibonacci Generator sometimes just called an LFG, is a particular type of pseudo random number generator. It is based on the Fibonacci sequence.

PIKE

This algorithm was published by Ross Anderson. It was proposed as an improvement on FISH. PIKE is both faster and more secure than FISH. The name PIKE is not an acronym, but rather a humorous play on the previous FISH, a pike being a type of fish.

Implementing Ciphers

The preceding section explored some widely used symmetric ciphers. That coverage was by no means exhaustive. There are many other symmetric ciphers and each of those ciphers could have been covered in more detail. However, the ciphers covered are those most widely used. The goal of this chapter is to provide you with a general familiarity with cryptography, not an exhaustive understanding. In this section methods of implementing block ciphers will be described.

Electronic Codebook

The most basic encryption mode is the electronic codebook (ECB) mode. This essentially means use the algorithm precisely as it is described with no alterations. This is quite common, but as will be shown, is not usually the right choice.

Cipher-Block Chaining

When using cipher-block chaining (CBC) mode, each block of plaintext is XORed with the previous ciphertext block before being encrypted. This completely negates a wide range of attacks, including known plain text attacks. Even if a message consisted of all blocks of the same exact plaintext, each cipher text block would be different. It should also be noted that the XOR operation is very fast and there should be no noticeable performance issues when implementing CBC.

The major concern with CBC regards the first block. There is no preceding block of cipher text to XOR the first plaintext block with. To overcome this issue, it is common to add an initialization vector to the first block so that it has something to be XORd with. The initialization vector is basically a pseudo random number, much like the cipher key. Usually an IV is only used once and is thus called a *nonce* (number used only once).

CFB and OFB

Cipher Feedback Mode (CFB) is related to CBC in structure but has a very different purpose. CFB is used to convert a block cipher into a stream cipher. Specifically, a self-synchronizing stream cipher. Output Feedback Mode (OFB) also takes a block cipher and makes it into a stream cipher. In this case it makes a synchronous stream cipher.

GCM

Galois/Counter Mode (GCM) is a mode used with block ciphers to provide data authentication as well as encryption. Block ciphers, by themselves, do not include authentication. GCM does provide authentication.

Cryptographic Hashes

A cryptographic hash is a type of algorithm that has three primary properties. First it is a one-way function. There is no mechanism to reverse a hash. The second property is fixed length output. Each hashing algorithm puts out a specific size of output, regardless of the size of the input. Finally, a hash must be collision resistant. If you put in two different inputs and get out the same output, that is a collision. This is shown in Fig. 1.3.

The reason this is even possible is because each hashing algorithm has a fixed length output. For example, MD5 has a 128-bit output. That means they are only 2^{128} power possible outputs. If you take $2^{128} + 1$ different inputs, then the last output must match one of the previous outputs.

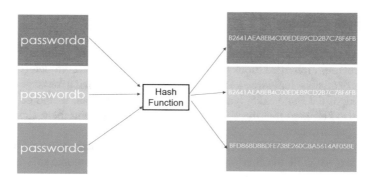

Fig. 1.3 Hash collision

MD5

This is a 128-bit hash designed by Ron Rivest in 1991. MD5 is specified by RFC 1321. As early as 1996 a flaw as found in MD5 and by 2004 it was shown that MD5 was not collision resistant. MD5 is no longer considered secure and should not be used.

SHA

The Secure Hash Algorithm is actually a family of hashing algorithms. This is also perhaps the most widely used hash algorithm today. There are now several versions of SHA. SHA (all versions) is considered secure and collision free.

> SHA-1: This is a 160-bit hash function which resembles the earlier MD5 algorithm. This was designed by the National Security Agency (NSA) to be part of the Digital Signature Algorithm.
>
> SHA-2: This is actually two similar hash functions, with different block sizes, known as SHA-256 and SHA-512. They differ in the word size; SHA-256 uses 32-byte (256 bits) words where SHA-512 uses 64-byte (512 bits) words. There are also truncated versions of each standardized, known as SHA-224 and SHA-384. These were also designed by the NSA.
>
> SHA-3: This is the latest version of SHA. It was adopted in October of 2012 when the Keccak algorithm was chosen to be SHA-3.

RipeMD

RipeMD is not as well-known as SHA but is still a prominent hashing algorithm. RACE Integrity Primitives Evaluation Message Digest is a 160-bit hash algorithm developed by Hans Dobbertin, Antoon Bosselaers and Bart Preneel. There exist 128, 256 and 320-bit versions of this algorithm. This algorithm was developed in Europe and is recommended by the German government.

HAVAL

HAVAL is a cryptographic hash function. HAVAL can produce hashes of different lengths. HAVAL can produce hashes in lengths of 128 bits, 160 bits, 192 bits, 224 bits, and 256 bits. HAVAL was published in 1992 by Yuliang Zheng, Josef Pieprzyk, and Jennifer Seberry.

Tiger

Tiger is hashing algorithm published by Ross Anderson and Eli Biham, in 1995. Tiger produces a 192-bit digest. This algorithm works well in software implementations but is not efficient in hardware implementations.

MAC and HMAC

Hashes are used for many purposes, as was explained earlier. However, there is a weakness to hashing. If you use a hash for message integrity, then it is possible for an attacker to intercept the message, delete your hash, change the message, then recompute the hash. Hashes, by themselves, do nothing to prevent this. Thus, Message Authentication Codes were developed. A Message Authentication Code (or MAC) incorporates a pre-shared key with hashing. There are two primary methods for accomplishing this.

One way to add keying to a hash is the HMAC or Hashing Message Authentication Code. The two parties pre-share a key of the same size as the digest that is output from their chosen hashing algorithm. For example, if two parties wish to use SHA-1 for message integrity, they first exchange a 160-bit key. Then each time a SHA-1 hash is calculated, it is then XOR'd with the 160-bit key. Thus, if the message is intercepted and the hash is changed, it will be detected on the receiving in.

Other option is to use a symmetric cipher in CBC (Cipher Block Chaining mode) and then use only the final block as the MAC. These are called CBC-MAC. Given that only the final block is used, the algorithm will have a fixed length output. Since a symmetric cipher is being used, there is a pre-shared key involved in the process.

Asymmetric Cryptography

In addition to symmetric ciphers and cryptographic hash functions, there is one more class of cryptographic primitives: asymmetric ciphers. These ciphers use one key to encrypt the message and a separate key to decrypt. This is shown in Fig. 1.4.

While asymmetric ciphers are always slower than symmetric ciphers, they have a very important advantage. There is no need to exchange a key in advance. If Alice wants to send a secure message to Bob, she uses Bob's public key. Anyone can get his public key and send him messages, but that public key will not decrypt those messages. Only the private key will. Thus, the issue of key exchange is no longer relevant.

Fig. 1.4 Asymmetric cryptography

RSA

RSA is perhaps the most widely used asymmetric algorithm and has been for many decades. This public key method was first published in 1977 by three mathematicians: Ron Rivest, Adi Shamir, and Len Adleman. The name RSA is derived from the first letter of each mathematician's last name. The math for this asymmetric algorithm is relatively simple so we will describe it here. This algorithm is so widely used, that it is given more attention and detail than other algorithms.

Before we can delve into RSA, there are a few basic math concepts you need to know. Some (or even all) of this material may be a review.

Prime Numbers: A prime number is divisible by itself and 1. So 2, 3, 5, 7, 11, 13, 17, 29, 31, etc. are all prime numbers. (Note that 1 itself is considered a special case and is not prime.)

Co-prime: This actually does not mean prime; it means two numbers have no common factors. So, for example, the factors of 8 (excluding the special case of 1) are 2 and 4. The factors of 9 are 3. The numbers 8 and 9 have no common factors. They are co-prime or this is sometimes called relatively prime.

Euhler's Totient: Pronounced "oilers" totient after the famous mathematician who originally published this, Leonard Euhler. This is how many numbers smaller than some given number (we will call n) have no common factors with n. Sometimes this is simply called the totient. Consider the number 10, the following table evaluates every number smaller than 10 to see if it has any common factors with 10:

Number	Any common factors
1	No (one is a special case)
2	Yes, 2 is a factor of 10
3	No
4	Yes, 2 is a factor of both 4 and 10
5	Yes, 5 is a factor of 10
6	Yes, 2 is a factor of both 6 and 10
7	No
8	Yes, 2 is a factor of both 8 and 10
9	No. The factors of 9 are 3; the factors of 10 are 2 and 5

Thus, there are four numbers (1, 3, 7, and 9) that are smaller than 10 and have no common factors with 10. Thus, we can say the Euhler's totient of 10 is 4. It is also the case that if you pick a prime number, the totient of that number is $n - 1$. The following table shows this with the number 5 (a prime number).

Number	Any common factors
1	No (one is a special case)
2	No
3	No
4	No

As you can see, there are 4 number that are smaller than 5 and have no common factors with 5. Thus, the Euhler's totient of 5 is 4.

Multiplying and co-prime: Now we can easily compute the totient of any number. And we know automatically that the totient of any prime number n is just $n - 1$. But what if we multiply two primes? The product is clearly not prime. Consider 5 * 2. 10 is not a prime number. And if we have very large prime numbers, going through the process we did earlier is simply not an option. Leonard Euhler proved that if you take to prime numbers and multiply them together, the totient of that product is found by multiplying the totients of the two prime numbers. So, if we return to 5 * 2 = 10, the totient of 5 is 4 and the totient of 2 is 1, so the totient of 10 should be 4 * 1, or just 4. Well if you recall earlier, we already saw that indeed the totient of 10 is 4.

Modulus: This is the final mathematical concept you need in order to understand RSA. The modulus operation is a way to work within a limited group and still do standard math such as addition and multiplication. Imagine for a moment that you had an artificial world consisting of the integers 0 through 5. In that world, you can do operations like 2 + 2, because the answer, 4, is still within that world. But can you do 3 * 3? The answer would be nine, which does not exist in this artificial world. Well you can do such operations, if you add on mod 5.

The modulus operator circumscribes operations by the limiting value (in this case 5). A more simplistic explanation is that it divides two numbers, but only returns the remainder. Therefore $3 * 3 = 9$, but 9 mod 5 = 4 and 4 is in the universe of integers from 0 to 5. In fact, we can work with quite large numbers. You can even do 44 as long as you do mod 5. Because 44 mod 5 is 1 which is in the universe of integers from 0 to 5.

These basic mathematical concepts are sufficient for you to understand the RSA algorithm. Obviously, there is more depth on any of these topics. But this is sufficient for you to understand the algorithm.

To create the key, you start by generating two large random primes, p and q, of approximately equal size. You need to pick two numbers so that when multiplied together the product will be the size you want (2048 bits, 4096 bits, etc.).

Now multiply p and q to get n.

Let $n = pq$

The next step is to multiply the Euler's totient for each of these primes in order to compute the Euler's prime for the product of the two primes

Let $m = (p - 1)(q - 1)$

Now we are going to select another number. We will call this number e. We want to pick e so that it is co-prime to m. Any number smaller than m and co prime to m can be selected. However, for ease of use, e is often itself a prime number.

Choose a small number e, co-prime to m.

There is just one more step in generating the keys. One must solve a basic algebra problem to find a number d that when multiplied by e and modulo m would yield a 1.

Find d, such that de mod $m \equiv 1$

Note we used $\equiv$ instead of $=$. This is the congruence symbol. Now you will publish e and n as the public key. Keep d as the secret key. To encrypt, you simply take your message raised to the e power and modulo n.

CipherText $= plaintext\ e$ mod n

To decrypt, you take the cipher text and raise it to the d power modulo n.

PlainText $= ciphertext\ d$ mod n

The letter e is for encrypt and d for decrypt. As you can see this is not particularly complex mathematics. It is likely you were quite familiar with the individual mathematical operations before reading this chapter. The key is the difficulty of factoring a number into its prime factors. In the case of a large n (2048 bits long, for example) it is simply not practical to factor that number into its prime factors. If that were not the case, then any attacker could simply take n from the public key and factor out p and q, then derive the rest of the key (i.e. d).

Diffie–Hellman

Diffie–Hellman was the first publicly described asymmetric algorithm, however it is really a key exchange protocol. You do not encrypt or decrypt messages with Diffie–Hellman. Rather you generate a key to be used. This is a cryptographic protocol that allows two parties to establish a shared key over an insecure channel. In other words, Diffie–Hellman is often used to allow parties to exchange a symmetric key through some unsecure medium, such as the Internet. It was developed by Whitfield Diffie and Martin Hellman in 1976. There have been modifications to Diffie–Hellman, most notably ElGamal and MQV.

ElGamal was first described by Taher Elgamal in 1984.ElGamal is based on the Diffie Hellman key exchange algorithm described earlier in this chapter. It is used in some versions of PGP. MQV (Menezes–Qu–Vanstone) is a protocol for key agreement that is based on Diffie–Hellman. It was first proposed by Menezes, Qu and Vanstone in 1995 then modified in 1998. MQV is incorporated in the public-key standard IEEE P1363.

Elliptic Curve

This algorithm was first described in 1985 by Victor Miller and Neil Koblitz. Elliptic Curve cryptography is based on the fact that finding the discrete logarithm of a random elliptic curve element with respect to a publicly known base point is difficult. By difficult, it is meant that it is impractical to accomplish. Elliptic Curve has been shown to be just as secure as RSA but with much smaller key sizes. For example, a 384-bit ECC key is as strong as 2048-bit RSA. There are a number of variations such as ECC-DH (ECC Diffie-Hellman), and ECC-DSA (ECC Digital Signature Algorithm).

Wireless Encryption and Security

In addition to understanding the algorithms used in cryptography, it is important to understand the applications of cryptography. One of the most common applications for cryptographic algorithms, is in the security of wireless communications.

WEP

Wired Equivalent Privacy uses the stream cipher RC4 to secure the data and a CRC-32 checksum for error checking. Standard WEP uses a 40-bit key (known as WEP-40) with a 24-bit initialization vector, to effectively form 64-bit encryption. 128-bit WEP uses a 104-bit key with a 24 bit IV.

Because RC4 is a stream cipher, the same traffic key must never be used twice. The purpose of an IV, which is transmitted as plain text, is to prevent any repetition, but a 24-bit IV is not long enough to ensure this on a busy network. The way the IV was used also opened WEP to a related key attack. For a 24-bit IV, there is a 50% probability the same IV will repeat after 5000 packets. WEP is insecure and should no longer be used.

WPA

WPA was a temporary measure to provide more security than WEP, while a solution that fully implemented the security requirements of 802.11 was finalized. Wi-Fi Protected Access. WPA uses Temporal Key Integrity Protocol. TKIP is a 128-bit per-packet key, meaning that it dynamically generates a new key for each packet.

WPA 2

WPA2 fully implements the IEEE 802.11i standard. It uses Advanced Encryption Standard (AES) using the Counter Mode-Cipher Block Chaining (CBC)-Message Authentication Code (MAC) Protocol (CCMP) that provides data confidentiality, data origin authentication, and data integrity for wireless frames.

WPA Enterprise provides RADIUS-based authentication using 802.1X. WPA Personal uses a pre-shared Shared Key (PSK) to establish the security using an 8 to 63-character passphrase. The PSK may also be entered as a 64-character hexadecimal string. It also uses EAP variations for authentication.

WPA 3

WPA3, was released in 2018. It has many new features. One prominent security feature is that WPA 3 requires attackers to interact with your Wi-Fi for every password guess they make, making it much harder and time-consuming to crack. Another important security feature is that with WPA3, even open networks will encrypt your individual traffic between the computer and the Wireless Access Point.

Obfuscation

While cryptography is primarily concerned with using mathematics to secure information, there are other methods for securing data. Information can be obfuscated so that it does not appear to be of value. The most common mechanism for doing this is steganography.

Steganography

Steganography seeks to hide information in some file. The most widely known form of steganography is to hide data in an image file. One takes a document, text file, or even another image file and hides it into an innocuous image file. Someone examining the image file cannot see from the image that anything is hidden in that file.

There are some basic steganography terms you should know before we continue:

Payload is the data to be covertly communicated. In other words, it is the message you wish to hide.
The *carrier* is the signal, stream, or data file into which the payload is hidden.
The *channel* is the type of medium used. This may be still photos, video, or sound files.

The most common way steganography and the method most often discussed in books is via least significant bits (LSB). In every file there are a certain number of bits per unit of the file. For example, an image file (such as a JPG, or BMP) in Windows has 24 bits per pixel. These are subdivided into 8 bits for red; 8 bits for green; and 8 bits for blue (RGB). If you change the least significant of those bits, then the change is not noticeable with the naked eye. If you change the lease significant bits of enough different pixels, you can hide data in that image.

That is only one way of hiding data. It is also possible to hide data in the distortions of music files. More complex mathematical functions such as the Discrete Cosine Transform are used to hide data in music or video.

There have also been historical methods of hiding data that predate modern computers. A few interesting examples are listed here:

The ancient Chinese wrapped notes in wax and swallowed them for transport. If the messenger was intercepted and searched, no message would be found.

In ancient Greece a messenger's head might be shaved, a message written on his head, then his hair was allowed to grow back.

In 1518 Johannes Trithmeus wrote a book on cryptography and described a technique where a message was hidden by having each letter taken as a word from a specific column.

These ancient methods illustrate the fact that the desire to hide messages has not changed, only the specific methodology has changed. Modern computers provide more methods to hide data.

Fig. 1.5 Invisible secrets

One popular tool for performing steganography is Invisible Secrets. The landing page for invisible secrets is shown in Fig. 1.5.

This tool allows you to encrypt files and hide them in other files. It also provides several options for the type of file you can hide data in, including JPG, HTML, and WAV files.

TOR

Onion routing is a method of hiding the source and destination address of a network packet. The packet is encrypted with layers of encryption (like an onion) and each intermediate router can only decrypt one layer. That will show that router the next place to send the packet. Should one intercept a packet in transit between two proxies, one can only determine the previous proxy and the next proxy. You cannot determine the actual origin or destination. This is shown in Fig. 1.6.

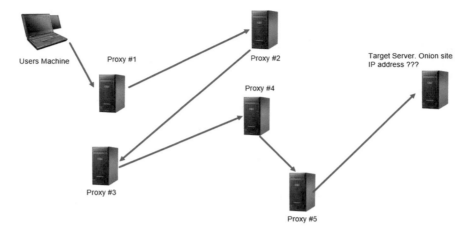

Fig. 1.6 The TOR network

This process of onion routing makes for a level an anonymity that is not readily available with traditional network traffic. While one can certainly utilize a fake identity on any website, the website may track the users IP address, thus revealing who the user is. It is certainly possible to spoof an IP address, or to use a public internet connection, but these only provide a small degree of anonymity. Onion routing makes the entire communication process anonymous.

This brings us to The Onion Router (TOR). This is a network that uses only onion routing. One of the most common ways to access this is to use the TOR browser. The TOR browser is a modified Firefox browser that can be downloaded for free.

Websites accessible only via TOR, make up what is commonly called the dark web. These sites usually end in a ".onion" address. For example, the address: http://kbhpodhnfxl3clb4.onion is a popular dark web search engine. These sites can only be accessed via the TOR browser. It is important to note that the TOR browser can be used to access normal web sites as well. However, dark web sites can only be accessed via the TOR browser.

Before we briefly discuss what is on the dark web it is important to note that the dark web is not merely for criminal activity. It provides a means for people to anonymously interact on the internet. It should be apparent to you that your internet traffic is highly monitored by many companies and search engines. Some people object to that. Thus, TOR is attractive to them. They may have no criminal intent at all. They just do not wish to be monitored.

With that said, there clearly is a great deal of criminal activity on the dark web. Of particular interest are dark web markets. These are markets that sell illegal goods and services. Payment is made in bit coin or similar currency. Stolen bank accounts, malware, drugs, and child pornography are rampant on dark web markets.

Fig. 1.7 Digital signatures

Digital Signatures

A digital signature is not used to ensure the confidentiality of a message but rather to guarantee who sent the message. This is referred to as nonrepudiation. Essentially, it proves who the sender is. Digital signatures are actually rather simple, but clever. They simply reverse the asymmetric encryption process. Recall that in asymmetric encryption, the public key (which anyone can have access to) is used to encrypt a message to the recipient, and the private key (which is kept secure, and private) can decrypt it. With a digital signature, the sender encrypts something with his private key. If the recipient is able to decrypt that with the sender's public key, then it must have been sent by the person purported to have sent the message (Fig. 1.7).

Remember, the concept is not confidentiality, but rather ensuring that the recipient can be confident in the identity of the sender. A good example is the digitally signing of printer drivers. The printer manufacturer has no need for confidentiality. In fact, they would probably be quite pleased if all consumers utilized their product and thus needed their drivers. However, their customers need a mechanism that will allow them to verify that the driver is truly from the manufacturer. This is how digital signatures are used.

Digital Certificates

In the discussion of public key cryptography, it was mentioned that the public key can be dispersed as widely as you wish, without diminishing security. Anyone who wishes to send you a message will encrypt it with your public key. As you saw in the discussion on RSA, someone else with that public key cannot decrypt the message. Only you, with your private key can do that. However, this does not address the question of how one would obtain another person's public key. This is typically done with a digital certificate. Digital certificates contain the person (or organizations) public key as well as information allowing you to confirm that person (or organizations) identity. The most common standard for digital certificates is the X.509 standard. All X.509 digital certificates have at least the following information:

- Version
- Certificate holder's public key info

  - Public Key Algorithm
  - Certificate holder's Public Key

- Serial number
- Certificate holder's distinguished name
- Certificate's validity period
- Unique name of certificate issuer
- Digital signature of issuer
- Signature algorithm identifier.

The version simply identifies the version of the X.509 standard that is being used in this certificate. This standard was first introduced in July of 1988. The current version is X.509 version 3, which is specified as a standard in RFC 5280. The next two items tell you what public key algorithm (RSA, Diffie–Hellman, Elliptic Curve, etc.) the certificate holder is using and give you the certificate holder's public key. Then there is a serial number to identify the certificate. Next, we get to the certificate holders distinguished name. By distinguished, it is meant unique. Such as a domain name or email address. Then there is the validity period. Normally certificates are issued for one year.

Now if this is all the digital certificate provided, then there would be no way to be certain that the person with this certificate is who they say they are. An attacker could easily forge a certificate claiming to be a bank, and setup a fake banking website. However, the other data in the X.509 certificate helps to mitigate that threat.

The unique name of the certificate issue tells who issued this certificate. Then the certificate includes the digital signature and algorithm used by that certificate issue. What usually is signed, is a hash of the contents of the certificate. So, when your browser encounters a website that is using a digital certificate, the first thing it does is verify the certificate issuers signature. If the certificate was indeed issued by whom it claims, then the next step is to check the data that was hashed. The browser will compare its own hash of the data, to that provided by the certificate issue. This way if any data has been modified, that will be detected. Then if that also checks out, the validity period is checked. All of these steps are done so that the recipient of the X.509 corticate can not only get the certificate holders public key, but also be assured that the certificate holder is who they claim to be.

There are also a number of terms associated with digital certificates. You should familiarize yourself with these terms:

- PKI (public key infrastructure) uses asymmetric key pairs and combines software, encryption and services to provide a means of protecting security of business communication and transactions.
- PKCS (Public Key Cryptography Standards) Put in place by RSA to ensure uniform Certificate management throughout the internet.
- A Certificate is a digital representation of information that identifies you as a relevant entity by a trusted third party (TTP).

- A CA (Certification Authority) is an entity trusted by one or more users to manage certificates.
- RA (Registration Authority) Used to take the burden off of a CA by handling verification prior to certificates being issued. RA acts as a proxy between user and CA. RA receives request, authenticates it and forwards it to the CA.
- CPA (Certificate Practice Statement) describes how the CA plans to manage the certificates it issues.
- CP (Certificate Policy) is a set of rules that defines how a certificate may be used.
- X.509 This is an international standard for the format and information contained in a digital certificate.
- CRL (Certificate Revocation List) is a list of certificates issued by a CA that are no longer valid. CRLs are distributed in two main ways: PUSH model: CA automatically sends the CRL out a regular interval. Pull model: The CRL is downloaded from the CA by those who want to see it to verify a certificate. End user is responsible.
- Status Checking: The newer method is the "Online Certificate Status Checking Protocol" called OCSP. It works in real time, unlike the certificate revocation list (CRL).

The preceding list of terms is critical to understanding digital certificates. These terms explain key functionality in digital certificates and the public key infrastructure (PKI).

SSL/TLS

The world wide web was born in the very early 1990s. It quickly became apparent that it would be desirable to be able to conduct financial transactions over the web. That required some process of encrypting and authenticating communications. The company Netscape originally developed Secure Sockets Layer (SSL). That was later replaced by Transport Layer Security (TLS). When you visit a website that has https at the beginning rather than http, the s indicates that site is using TLS to secure the traffic. Note, many sources still say SSL, but as you will see TLS was released in 1999, it is extremely doubtful many websites have not moved from SSL to TLS.

A very brief history of the development of SSL and TLS is provided below. Note that SSL became a public standard and has been governed as such for many years. It is no longer managed by Netscape.

- Unreleased v1 (Netscape)
- Version 2 released in 1995 but had many flaws
- Version 3 released in 1996 RFC 6101
- Standard TLS1.0 RFC 2246 released in 1999
- TLS 1.1 was defined in RFC 4346 in April 2006
- TLS 1.2 was defined in RFC 5246 in August 2008. It is based on the earlier TLS 1.1 spec
- TLS 1.3 was defined in RFC 8446 in August 2018 (Fig. 1.8).

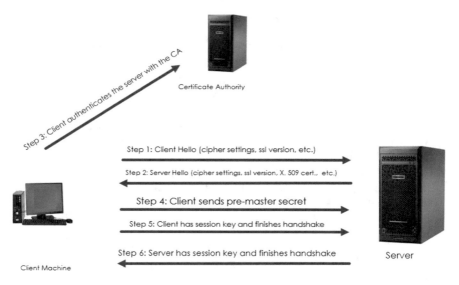

Fig. 1.8 SSL/TLS

The steps shown above are described in more detail in the following paragraphs.

1. The client sends the server information regarding the client's cryptographic capabilities. That includes what algorithms it is capable of, what hashing algorithms it can use for message integrity, and related information.
2. The server responds by selecting the best encryption and hashing that both client and server are capable of and sends this information to the client. The server also sends its own certificate, and if the client is requesting a server resource that requires client authentication, the server requests the client's certificate.
3. The client uses the information sent by the server to authenticate the server. This means authenticating the digital certificate with the appropriate certificate authority (CA). Originally this required communication with the certificate authority that issued the certificate. Modern computers come with certificates for all the major certificate authorities and can authenticate the digital signature without communicating with the CA. If this fails, the browser warns the user that the certificate cannot be verified. If the server can be successfully authenticated, the client proceeds to the next step. If the server has requested client authentication, then the server will also authenticate the client's X.509 certificate. This does not happen in most e-commerce and banking websites. This step is optional and not shown above
4. Using all data generated in the handshake thus far, the client creates the pre-master secret for the session, encrypts it with the server's public key that it received from the server's X.509 certificate, and then sends the encrypted pre-master secret to the server. Both the client and the server use the master secret to generate the session keys. These are symmetric keys (such as AES) that will

be used throughout the session to encrypt information between the client and the server.

5. The client sends a message to the server informing it that future messages from the client will be encrypted with the session key.

6. The server sends a message to the client informing it that future messages from the server will be encrypted with the session key.

This is the process your browser goes through every time you visit a website that uses HTTPS. So, every time you shop online, go to your bank, or do a myriad of other online activities, this process is used to secure the communications. Now the description of SSL/TLS shown in this section is still somewhat simplified but does capture all the essential elements of that process.

Cryptography Regulations

United States Regulations/Standards

The United States National Institute of Standards NIST) publishes a number of standards documents to cover the use of cryptography. FIPS (Federal Information Processing Standard) 197 covers the use of AES. Recall that AES was discussed in some detail earlier in this chapter. One of the earliest cryptographic regulations was NIST Special Publication (SP) 800-67 which defined DES as an encryption standard. Skipjack, the algorithm used in the Clipper chip was defined by FIPS 185. The Block Cipher modes such as ECB and CBC that were discussed earlier in this chapter are defined by FIPS 81.

Cryptography Laws

In general, as of this writing, in order to export cryptographic products from the United States to other nations, one needs a license from the U.S. Commerce Department. And, for the most part, these restrictions have been relaxed. For example, McAfee's Data Protection Suite, which includes encryption, has been granted an "ENC/Unrestricted" license exception by the U.S. Department of Commerce.

In the United States cryptography (as well as many other items) for use in government systems is regulated by Federal Information Processing Standard. Civilian use of cryptography is not regulated (except for the export of cryptography). The standards have relaxed in recent years. When exporting cryptography, the destination is important. There are three groups of countries. Group B are countries that are considered friendly and there are relaxed export rules. Group D:1 are under stricter rules. This group includes Russia and China. The third group, E:1 is a list of 5 countries considered terrorist supporting countries. And no cryptography can be exported to those countries.

Key Management

Storage and management of keys is a rather critical issue that must be addressed. The topic of key management encompasses setting up servers, establishing policies, generating keys, storing keys, and the destruction of keys when no longer needed. Of critical importance is the storage of keys. It is important to establish clear policies regarding key management.

Clearly, any server that is used to store cryptographic keys is a very important server. And it is a target for attackers. Thus, the security of that server is of paramount importance. This is an excellent place to implement the security concept of separation of duties. No one person should be able to retrieve keys from that server. In an ideal situation the server would not be networked and would be in a physically secure room. One person would have access to that room, but not have login credentials for the server. A second person would be able to login to the server, but not access an encrypted folder which holds the keys. A third person would have the key to that folder. This scenario requires three people (presumably trusted people who have been thoroughly vetted), to work in concert to retrieve a key from the key storage server. This mitigates the threat of an inside employee stealing other employees' keys from the key storage server.

A key management system (KMS) is an integrated approach to handling keys. This includes the generation of keys as well as the distribution of keys to devices that require them. There is a key management interoperability protocol (KMIP) that was developed to facility exchange of information between diverse systems. Message formats are defined. KMIP supports the following operations:

- Create—Create a key.
- Get—Retrieve a key from the system.
- Register—Add an externally generated key to the system.
- Add Attributes—Add an attribute to an object.
- Get Attributes—Retrieve an objects attribute.
- Modify Attributes—Modify an objects attribute.
- Locate—Retrieve objects.
- Re-Key—Regenerate key pairs.
- Create Key Pair—Generate key pairs.
- Certify—Certify a digital certificate.
- Encrypt—Encrypt using a key.

- Decrypt—Decrypt using a key.
- Import—Import an externally generated key (note: you first import, then register it).
- Export—Export a key for use in some other system.

These operations are the essential ones required by any key management system. They are part of the message structure for KMIP. This protocol facilitates key management.

There are also standards that provide guidance on key management processes and procedures. U.S. National Institute of Standards (NIST) Special Publication (SP) 800-57 provides guidance on key management. Part 1 handles general guidelines. Part 2 discusses key management organizations. Then finally part 3 covers application specific key management guidance.

Drive and File Encryption

One of the most essential things one can do with cryptography, is to encrypt sensitive data. There are two approaches to this. The first is to encrypt individual files separately. The second is to encrypt the entire drive. Each approach has its own merits and issues.

Full drive encryption is an excellent way to protect data should the device become lost or stolen. It prevents an attacker from accessing anything on that computer. Furthermore, drive encryption mitigates many Windows password hacking attacks. However, to use the device, you must obviously decrypt the drive. During all operational hours, the device will be decrypted. This provides a myriad of options for an attacker to attempt to exfiltrate data. As one example, spyware on the device would be able to access all the files on the device. This is one reason it is often a good idea to also encrypt specific, highly sensitive files.

XTS-AES is a technique for improving hard drive encryption. XTS-AES is standardized as IEEE P1619. It supports using a different key for the initialization vector than for the block cipher.

File encryption is the encryption of a single file, group of files, or a folder. This is separate from drive encryption. Even should one gain complete access to the device, encrypted files would still be protected. It is recommended that you combine file and drive encryption for a more complete security posture.

Microsoft Bitlocker is a drive encryption approach that is part of Windows and readily available in most Windows versions. This makes it an attractive solution for Windows users. It is also very easy to use. You can see the Bitlocker control interface in Fig. 1.9.

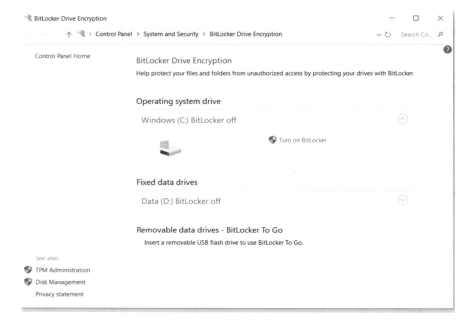

Fig. 1.9 Microsoft Bitlocker

Virtual Private Networks

A *VPN* is a *virtual private network*. The concept is to provide the remote user with the same, or substantially similar access to what he or she would have if they were on the organizations network.

There are four different protocols that are used to create VPNs:

- Point-to-Point Tunneling Protocol (PPTP)
- Layer 2 Tunneling Protocol (L2TP)
- Internet Protocol Security (IPsec)
- SSL/TLS.

These are presented in historical order from the oldest to the most recent. However, it should be noted that even the older protocols are still in use.

These are each discussed in more depth in the following sections.

Point-to-Point Tunneling Protocol

Point-to-Point Tunneling Protocol (PPTP) is the oldest of the three protocols used in VPNs. It was originally designed as a secure extension to Point-to-Point Protocol

(PPP). It essentially adds security to the PPP protocol. PPP was meant to connect two devices but had no security mechanisms. PPTP was originally proposed as a standard RFC 2637 in 1999. The protocol adds the features of encrypting packets and authenticating users to the older PPP protocol. PPTP works at the data link layer of the OSI model.

PPTP creates a connection using TCP port 1723. That connection is used to initiate and manage a tunnel. The tunnel is created using the GRE (Generic Routing Encapsulation) protocol. PPTP offers two different methods of authenticating the user: Extensible Authentication Protocol (EAP) and Challenge Handshake Authentication Protocol (CHAP). EAP was actually designed specifically for PPTP and is not proprietary. CHAP is a three-way process whereby the client sends a code to the server, the server authenticates it, and then the server responds to the client. CHAP also periodically reauthenticates a remote client, even after the connection is established.

Layer 2 Tunneling Protocol

Layer 2 Tunneling Protocol (L2TP) was designed as an improvement to PPTP. Like PPTP, it works at the data link layer of the OSI model. It has several improvements to PPTP. First, it offers more and varied methods for authentication PPTP offers two, whereas L2TP offers five. In addition to CHAP and EAP, L2TP offers PAP, SPAP, and MS-CHAP.

Besides more authentication protocols available for use, L2TP offers other enhancements. PPTP will only work over standard IP networks, whereas L2TP will work over X.25 networks (a common protocol in phone systems) and ATM (asynchronous transfer mode, a high-speed networking technology) systems. L2TP also uses IPsec for its encryption.

IPsec

IPsec is currently the most widely used of the VPN protocols. IPSec operates in one of two modes: Transport mode, in which only the payload is encrypted, and Tunnel mode, in which both data and IP headers are encrypted. Following are some basic IPsec terms:

Authentication Headers (AHs) provide connectionless integrity and data origin authentication for IP packets.

Encapsulating Security Payloads (ESPs) provide origin authenticity, integrity, and confidentiality protection of packets. These have encryption-only and authentication-only configurations.

Security Associations (SAs) provide the parameters necessary for AH or ESP operations. SAs are established using the Internet Security Association and Key Management Protocol.

The Internet Security Association and Key Management Protocol (ISAKMP) provides a framework for authentication and key exchange.

Internet key exchange (IKE and IKEv2) is used to set up a security association (SA) by handling negotiation of protocols and algorithms and to generate the encryption and authentication keys to be used.

Essentially during the initial establishment of an IPsec tunnel, security associations (SAs) are formed. These SAs have information such as what encryption algorithm and what hashing algorithms will be used in the IPsec tunnel. IKE is primarily concerned with establishing these SAs. ISAKMP allows the two ends of the IPsec tunnel to authenticate to each other and to exchange keys.

SSL/TLS

SSL/TLS VPN's are newer than the other three protocols. They initiate a VPN connection in much the same was a TLS is normally done. However, once a connection is made, the user has access to the network, rather than just access to some data on a web page. Essentially the SSL/TLS network uses the same technology described earlier in this chapter to secure web pages, but instead uses it to initiate a VPN.

Conclusion

This chapter began with a discussion of ancient ciphers. Then modern symmetric, asymmetric, and hashing algorithms were discussed. A wide range of applications of cryptography including Wi-Fi security, SSL/TLS, and VPN's where discussed. It is important that you master the topics in this chapter before proceeding. The elementary concepts of cryptography presented here are fundamental to a wide range of security topics you will encounter later in this book.

Chapter 2
Information Systems Security Management

Izzat Alsmadi

Introduction

Information Systems Security (ISS) is a broad term that covers all security aspects in enterprises, including system analysis and design method, manual information systems, managerial issues, and both societal and ethical problems.

Information systems security management (ISSM) evolved recently to be a discipline by itself. Some Universities offer graduate programs in ISSM. It integrates general information systems skills with security and management skills. One of ISSM major goals is to ensure the confidentiality, integrity, and availability of information whenever and wherever it is transmitted, stored, and processed.

ISSM Covers Two Dimensions from Information Systems' Perspectives:

- The ability to consider all security aspects in all enterprise information systems. For example, an Information systems security (ISS) manager works as a Chief Security Officer (CSO) to defend all information systems and be in charge of tasks related to security analysis, design, and implementation in all information systems.
- As an alternative, ISSM indicates the need to view security itself as an information system and apply known system analysis, design, and management principles in acquiring and management security systems or controls.

K0005: Knowledge of Cyber Threats and Vulnerabilities

This knowledge competency is one of the six core knowledge areas in NICE framework, Table 2.1.

Those are included in all 52 specialty areas (33 main specialty areas, with several sub-specialty areas). Looking at the description of those Knowledge competencies,

© The Editor(s) (if applicable) and The Author(s), under exclusive license to Springer Nature Switzerland AG 2020
I. Alsmadi et al., *The NICE Cyber Security Framework*,
https://doi.org/10.1007/978-3-030-41987-5_2

Table 2.1 NICE 6 core
knowledge areas for all work
roles

K0001	K0004
K0002	K0005
K0003	K0006

we can see that they are very broad in nature. Each one of them can be covered within a section, a chapter, or a course.

Cyber threats, and related vulnerabilities, are becoming more and more sophisticated. Eventually, the job of keeping defense mechanisms one step ahead is a very challenging task.

In cyber threats, we focus on studying sources of attacks. Through the analysis and understanding of threats, security policies, and procedures can be created to protect against certain those types of cyber-attacks or threats. Vulnerabilities on the other hand, focus on studying our own information systems, assets, applications, users, policies, etc. looking for weaknesses that can trigger such attacks. Ultimately, security team goal is to eliminate all existing or discovered vulnerabilities. Practically, as some vulnerabilities are impractical or expensive to accommodate for, or to accommodate for undiscovered vulnerabilities, security team should also consider techniques to tolerate some vulnerabilities or reduce the impact once those triggered.

Cyber defenders need to have skills in both sides to be able to protect their systems, the ability to know how to test their own systems for possible vulnerabilities and the knowledge of the different types of threats in their environment and the types of security controls that can be used to protect against them.

Cyber Threat Categories

The types of cyber security threats evolved rapidly with major technology trends such as the Internet, smart phones, Online Social Networks (OSNs), IoT, etc. Table 2.2 shows some of the popular categories of security threats over history.

We will cover those briefly in the next section. Readers are recommended to read more details in relevant references.

1. *Viruses*

Table 2.2 Most common
cyber security threats (over
history)

Viruses	DoS attacks
Worms	Phishing attacks
Rogue security software	SQL injection
Backdoors, Trojan Horses, and Rootkits	XSS
Adwares and spywares	Ransomwares

Viruses represent the oldest and most classical type of malwares or attacks before the era of the Internet. They are the only categories that do not need the Internet or related services/applications to propagate from one user to another or from one computing system to another and infect new files or applications. Their host that carry them are files or applications such as Microsoft Office (i.e., Word, Excel, etc.) files or applications. They can spread using floppy disks, CDs, DVDs, USB drives, etc. They are very less popular in these days due to the obsolescence of their medias of propagations and the ability of most anti-malware systems to detect and eliminate them.

There are three major characteristics that can uniquely identify a certain virus type, or any malware in general:

1. Infection method: How such virus injects its payload or reach its victims.
2. Propagation method: How the virus moves from one victim machine or file to another.
3. Payload: The payload is probably the most popular attribute that is used to uniquely identify and classify viruses. We will use those three characteristics when we discuss details on the different types of viruses and malwares.

Next, we will cover categories of the most popular viruses.

1. File Infector

File infector viruses were very popular; they infect certain file types and attach to them and propagate from one file to another whenever the infected file is triggered or accessed.

Many of file infector viruses are memory-resident; once they have been executed, they remain active in the computer's memory and possibly infect other programs in the memory.

Files that are the most vulnerable to this type of infection have the extensions of EXE. (i.e., executables in Windows environment) and .COM (command), though any file that is capable of execution can be infected. Some of the popular File infector viruses include: Jerusalem, Win32.Sality.BK, Cascade, and Cleevix.

2. Boot Sector Viruses

Boot sector viruses were popular when floppy disks were used for booting a computer. If a computer is infected with a boot sector virus, it will run its operations whenever the operating system starts. Recently boot sector viruses are seen as a mean to install Bootkits. Examples of early popular boot sector viruses include: Yale, Ping Pong, Form, Disk Killer, Michelangelo, and Stoned.

3. MBR (Master Boot Record) Viruses

Master Boot Record (MBR) is located at the very first beginning or sector of the hard-disk.

Master boot record viruses reside in memory, similar to boot sector viruses.

The difference between boot sector and MBR viruses is where the viral code is located. MBR virus usually save a copy of the MBR in a different location. Examples of MBR viruses include: New York Boot (NYB), AntiExe, Polyboot.B, and Unashamed.

4. Macro Viruses

This is another type of file-based viruses. It attaches to files made from programs that support macros (e.g., Microsoft Excel and Power Point). Macro viruses came when most of existing viruses were using some type of executable, but Macro viruses were an exception; an Excel file can carry a virus to you. Examples of macro-viruses include W97M.Melissa, ILOVEYOU, WM.NiceDay and W97M.Groov, DMV, Nuclear O97M/Y2K, Bablas, and Relax.

5. Polymorphic Viruses

These viruses are harder to detect. This is because security programs scan coding to identify viruses, but these specific viruses both encrypt and change their coding. They continually changing their operations over time, which may affect the programs you use. The "Dark Avenger's Mutation Engine" (DAME also known as MTE) has been released by virus writers to add this capability to any virus. Examples of Polymorphic Viruses: Satan Bug, Elkern, Tuareg, and Marburg.

6. Multipartite Viruses

Some types of computer viruses only spread in one way, but some others spread in several ways. Their actions usually depend on the form of operating system, certain programs on the computer or the usage of specific files or applications. It can have multiple actions, so it is complex and difficult to detect. For example, a Multipartite virus combines the features of a boot sector virus and a file infector virus. This strategy requires the virus to include a metamorphic engine, making it large and complex, but also very difficult to detect. Examples of Metamorphic viruses include: Involuntary, Stimulate, Cascade, Phoenix, Evil, One_Half, Emperor, Anthrax and Tequilla, Proud, Virus 101, Flip, Invader, and Win95.Zmist.A.

7. Direct Action or Non-resident Viruses

This type of viruses will be only active upon execution of the file or program it is attached to. Once the program is not in use, the virus stays in a dormant and no longer runs its operations. One of the early examples in this category are Virdem and Vienna viruses.

8. Browser Hijacker viruses

Those viruses cause original pages to be hijacked and users will keep bouncing between different pages. Some other malwares use this technique also (e.g., Phishing, MiM, etc.). Examples of those viruses: "Search the web", Playbar.biz or Quickneasysearch.com.

9. Stealth and Self-modifying Viruses

Some of the virus types earlier (e.g., AntiExe) use stealth techniques to hide themselves. Their main goal is to avoid or complicate detection. One technique is to ensure that file size goes back to its original size, despite the addition of malicious code to the file. Thus, it nullifies the ability to use the file length as an indicator of infection. At the end, almost all viruses include a degree of stealth as they attempt to conceal their presence in order to maximize their chances of spreading.

Self-modifying viruses are another kind of stealth viruses. Their goal is to internally change their signature to confuse signature-based malware detection tools and avoid detection.

Examples of stealth viruses include: Frodo, Joshi, and Whale.

10. Web Scripting Viruses

A site may unknowingly host malicious codes added by a third party. Those viruses breach the web browser security and allow the attackers to inject client-side scripting into the web page.

Generally, there are two different types of web scripting viruses: Temporary or non-persistent and persistent attacks that can steal web cookies and hijack users' sessions.

One example of this category is JS.Fortnight.

11. Memory Resident Viruses

This form of computer viruses embeds itself in the computer's memory to carry out real time operations. This means that whenever operating system is running, the virus is working. It may have different effects. It may clear up space on your computer for its own use by corrupting and deleting system files.

Examples of memory resident viruses include: CMJ, Meve, Randex, and Mrklunky.

12. FAT Viruses

FAT viruses target file allocation systems (e.g., FAT file systems). It may destroy files and the entire directories that hold them.

The link virus is one example of FAT viruses.

13. Spacefiller or Cavity Viruses

Spacefillers attach themselves to the file and can alter the start of the program. One example of this type is the Lehigh virus.

2. *DoS Attacks*

In Denial of Service (DoS) attacks, attackers try to deny legitimate users access target service through simulating a large number of service requestors and temporary or permanently interrupt such service. Despite the fact that this attack is very classical, yet it survives all those years and still range in top in terms in terms of types and

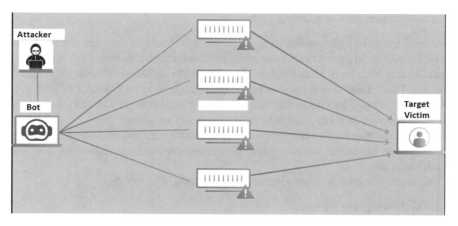

Fig. 2.1 DDoS example

volumes of attacks worldwide. Some of the reasons for such survival are the simplicity to craft DoS attacks. Additionally, from a defense perspective, many defense mechanisms can produce false negatives and positives, confusing legitimate service requests with attack requests.

If the incoming traffic comes from different sources, it is then called Distributed DoS (DDoS) where one or many hackers will synchronize their effort to accelerate and maximize the chance to bring the service down.

In many cases, machines that are used to start the DDoS attack are victims themselves. This is when hackers start their attack by controlling intermediate victim machines, called Zombies or Botnets. Those Zombies will the source of the second round of the attack, Fig. 2.1.

In addition to DoS and DDoS, another DoS attack category is called application or layer 7 DoS attack. A "Banana attack" such as "LAND" can fall under this category. In this attack, redirecting user outgoing messages back to them.

In one DDoS examples, DD4BC group launched in 2015 over 140 DDoS attacks on Akamai.

Targeting the availability of a website can have impacts related to the website reputation and credibility. In addition, in the e-commerce world, targeting a website that offer selling buying online can also mean financial loss. In addition to those goals, DoS can be the attack final goal when hackers are activists (e.g., for political, environmental, etc. reasons). As a result, the attackers' goal is to bring the target website down and maybe deface it with their own messages.

The first publicly recorded DoS attack was in 1997 during a DEF CON event. GitHub website was a victim of a large DoS attack in 2018.

In many attacks, DoS itself is not a goal but a tool to disrupt target services and further control or hack those services. Starting in 2000, Computer Emergency Response Team (CERT) warned that an increasing number of DoS attacks had begun to use DNS as a mean of amplifying bandwidth.

Bots and Botnets

Below are some examples of popular Botnets

- The first botnet to gain public infamy was a spammer built by Khan K. Smith in 2000.
- The Bredolab or Oficla botnet, possibly has the all-time record of controlled machines, was a Russian botnet mostly involved in e-mail spams. It was estimated that the botnet consists of 30 millions of zombie or bot computers.
- One major example of a Botnet was Zeus (Known also as Zbot); a Trojan horse that runs on MS Windows. One of the main usages of Zeus was to steal banking information using MiM techniques such as Spying methods (e.g., Browser keystroke logging and form grabbing). It became as the largest Botnet in the world with more than 3.5 million computers under its control. Zeus has also different variants such as Sphinx.
- In 2007, Storm peer-to-peer botnet was among the first to be controlled by several different servers.
- In 2008, Kraken botnet has 495,000 bots and was able to send as many as 600,000 emails per day.
- In 2009, the spam botnet Cutwail was sending millions of emails every minute.
- Andromeda or Gamarue was associated with the distribution of over 80 malware types.
- In 2016, Mirai botnet left the majority of US east coast with no Internet services. Mirai is a self-propagating botnet virus that can infect ill protected internet devices using Telnet.
- Koobface worm and botnet targeted many social networking sites such as Facebook, Twitter, and Myspace. It has been reported that Koobface has infected more than 3 million computers around the world.
- Monkif is known to download and install browser helper objects which can compromise web browsers.
- TR/Dldr.Agent.JKH is a botnet that once installed, remains in the system lying dormant until it receives commands from the controlling domain.
- An adware-serving botnet connecting Skype communications platform has been discovered in 2015.
- Swizzor is a botnet that can download rogue security applications and Trojans silently without users' notifications.

Flooding Attacks

In flooding attacks, attackers try to overwhelm users and their machines with high volumes of traffic. Flooding is a DoS attack category with main goal of disrupting services or bringing them down. Any entity with a public IP address can be a subject to flooding attacks. Security controls that try to block flooding attacks try to limit large size traffics. Some security roles to block flooding attacks may cause false positive cases when they identify legitimate traffic as malicious, if such traffic or applications include large volume or size of network traffic.

Countering flooding attacks may occur at the end hosts or at the network-level. End-host counter flooding techniques include:

- Increasing TCP Backlog
- Reducing the SYN-RECEIVED Timer
- Using SYN caches and SYN cookies

Network-based flooding counter measures include:

- Using IP, port, protocol filtering
- Using firewalls, access lists, proxies, and other types of access controls

In one classification, flooding attacks can be divided into three categories:

- Flooding based on protocol manipulations: Some of the popular types in this category include: SYN floods, fragmented packet attacks, Smurf attack, Ping of Death, etc.

A SYN flood occurs when a host is overwhelmed by SYN segments initiating incomplete connection requests.

- Volume-based flooding: Spoofed packets flooding such as ICMP or Ping and UDP flooding can fall under this category.
- Application-layer attacks: Attacks that target specific systems or applications (e.g., Windows, Apache), such as Get/Post attacks with main goal to crash those systems or applications.

3. *Worms and Phishing Attacks*

With the rise of the Internet, Worms because one of the most serious categories of malwares as they can infiltrate a large number of machines through the Internet within relatively a short amount of time. This is one of the main reasons why such malwares are called worms. Unlike viruses that use files as their hosts, worms can be standalones or can use the computing machines or systems as their hosts. Their key intelligence is on how to propagate fast through the different machines. Starting from one victim machines, they can then propagate to other IP addresses in the victim machine network, they can use email lists to propagate from one user email to all user friends in their address book. More recently worms start taking advantages of Online Social Networks (OSNs) accounts such as Facebook, Twitter, etc. to propagate to users' friends' lists.

Phishing in (Ph) rather than fishing, is a cyber attack related to tricking users to eat a bait. Typical baits can be emails or messages with a phishing message and a link (Fig. 2.2).

4. *Rogue Security Software*

Many scammers try to make money by creating fake malwares or fake messages that warn users for possible malwares in their systems. In reality, scammers use those messages to scare users and trick them to buy fake cleaning software or follow links that may eventually install the actual malicious software.

Information Regarding Your account:
Dear PayPal Member:

Attention! Your PayPal account has been limited!

As part of our security measures, we regularly screen activity in the PayPal system.We recently contacted you after noticing an issue on your account.We requested information from you for the following reason:

Our system detected unusual charges to a credit card linked to your PayPal account.

Reference Number: PP-259-187-991

This is the Last reminder to log in to PayPal as soon as possible. Once you log in, you will be provided with steps to restore your account access.

Once you log in, you will be provided with steps to restore your account access. We appreciate your understanding as we work to ensure account safety.

Click here to activate your account

Protect Your Account Info

Make sure you never provide your password to fraudulent websites.

To safely and securely access the PayPal website or your account, open a new web browser (e.g. Internet Explorer or Netscape) and type in the PayPal login page (http://paypal.com/) to be sure you are on the real PayPal site.

For more information on protecting yourself from fraud, please review our Security Tips at https://www.paypal.com/us/securitytips

Protect Your Password

You should never give your PayPal password to anyone.

Fig. 2.2 A sample of a phishing email

5. *Backdoors, Trojan Horses, and Rootkits*

Backdoors are examples of malicious applications that create unauthorized access to victim machines. Those backdoors can be part of another category of malwares that starts with or includes the need for a backdoor to allow root or administrator access.

Trojan horses imply using malwares that pretend to be applications with good or naive intentions. In the general definition, most malwares may disguise or embed themselves within good or known useful applications.

Rootkits are examples of tools or applications that try to have root access to system sensitive resources.

6. *SQL Injection*

In SQL injection attacks, attackers target webforms with malformed SQL queries to be able to access system backend databases. They use their knowledge of how SQL queries are formed on web portals front ends, transferred, and executed on the back-end databases. They may intend to cause destruction to some of the database components, data manipulation or just be able to expose and access records and tables in those databases.

One of the most important mechanisms to counter SQL injection attacks is to ensure proper user input validation techniques at the web front end or user interface level. Without being properly validated, those queries should not be sent or executed in the backend databases.

7. *XSS*

In Cross Site Scripting (XSS), attackers use malicious scripts (e.g., Java scripts) to expose victim web pages. Typically, the attacker doesn't attack the victim machine or web browser directly; rather, the attacker exploits a vulnerability in a website. Then the victim visits and gets the website to deliver unintentionally the malicious script for the attacker.

There are three main categories of cross-site scripting vulnerabilities: stored XSS (i.e., stored permanently on the target application), reflected XSS (i.e., the attacker's payload is part of the request sent to the web server and reflected back) and DOM-based XSS, when web application's client-side script writes user-provided data to the Document Object Model (DOM).

Input validation is also an important mechanism to mitigate against XSS. Certain inputs that may imply using scripts should be prevented.

8. *Adwares and Spywares*

Adwares are malicious applications that generate advertisements whenever the user is online. One major category of Adwares is: Pop-ups. They can be very annoying and users may struggle trying to get rid of them as in most cases, the installation/uninstallation mechanisms are hidden. In other cases, those Adwares hide in registry mechanisms to reinstall themselves whenever users try to delete them.

Spywares involve malicious software to spy on users' activities. One popular category of Spywares is Keyloggers that track and record all keys types by users on their machine to save them and email them to attackers. They may also include capturing mouse movements and user interface activities.

9. *Ransomwares*

Recently, Ransomwares start to get more popularity and pose serious large-scale threats. Examples of some of the very recent popular serious Ransomwares are: WannaCry and GandCrab. Many experts in the field believe that Ransomware attacks will continue to grow and be very serious and damaging.

As the name implies, in Ransomware attacks, attackers encrypt victim machine systems or files with encryption keys and demand Ransomwares to decrypt or release those files.

10. *Command & Control (C&C)*

Similar to the idea of backdoors in C&C, attackers deploy C7C servers on victim machines to be able to control them remotely. C&C are largely used by BotNets that we described earlier.

11. *Mobile Malware*

Recently, mobile malwares witness a significant rise, no wonder as users' usage of smart phones is growing rapidly.

Malware threats such as viruses, worms, spywares, Trojan horses, etc. typically target desktop and web platforms. Malwares in generally can be distinguished based on several factors:

- Access or intrusion method: Different malwares have their different ways of getting to victim machines. They can trick users to install them through masquerading as genuine applications. They may also fake their identities to look like genuine applications or embed themselves in innocent files. Users can be also tricked to install those malwares through different social engineering methods. For smart devices in particular, downloading some free applications from un-trusted application stores seem also to be a popular access method for those malwares.
- Propagation: Different malwares have their different propagation or expansion methods. The worst and most serious in terms of propagation are worms and spams as their main goal is to expand to the largest number of possible machines, smart devices or users. This expansion can be triggered through accounts (i.e., victim friends' list) in their accounts or email applications. They can be also triggered through certain networking schemes (i.e., expand gradually through local networks of victims). They do not require, usually, users' intervention to make the propagation trigger.
- Payload: Malwares are malicious applications developed usually to cause harm to victim machines. Such harm or payload can vary from one malware to another. Many complex malwares can have many types or categories of payloads. For example, the main payload of a spyware is to steel users' private information, financial data and possibly to commit identity theft. The main goal of an adware is to keep post unsolicited ads and links to users. The goal of mobile auto-rooters is to gain "root" level access privileges. This may usually lead to other types of payloads.
- Hiding methods: Recent advanced and complex malwares employ methods to hide, avoid detection, and avoid analysis or reverse engineering analysis. They typically will use some encryption or obfuscation method to defeat one or more of those previously mentioned counter attack methods. They may also hide or delay their payload to avoid detection. They may also act or behave differently in different time, environments, etc. to avoid or complicate detection methods.

Usage of Smart devices as a platform is growing rapidly in number of users as well as applications. While large scale mobile malware infections are insignificant yet, this is expected to change in future. Both mobile manufacturers and users should have the vision to expect that this may happen and should be prepared for it. Followings are examples of recent mobile malwares:

1. The Pegasus spyware: Spywares try to steal different types of information based on the nature of the attacks (e.g., identity theft, private data, pictures, listen to your audios, screen, etc.).

One of the most recent mobile malwares is Pegasus spyware. The malware is traced to a malware company in Israel (NSO Group) owned by a private US equity. The malware was discovered and Apple iOS released 9.3.5 to include required security updates within 10 days from the first time a human right defender received a message related to the malware. The message was forwarded and analyzed by Lookout security company.

Pegasus focuses on utilizing tools that come with most smart devices (e.g., WiFi, Camera, Microphones) to spy on users' videos, audios, messages, accounts, etc. The malware uses also encryption to avoid being easily detected. The malware analysis showed that it exploits three types of zero-day vulnerability in Apple iOS TM:

- CVE-2016-4655 (https://web.nvd.nist.gov/view/vuln/detail?vulnId=CVE-2016-4655).
- Kernel Information leak in Kernel. Apple iOS TM (before 9.3.5) allows attackers to obtain sensitive information from memory. A kernel base mapping vulnerability exists that leaks information allowing the attacker to calculate the kernel's location in memory.
- CVE-2016-4656 (https://web.nvd.nist.gov/view/vuln/detail?vulnId=CVE-2016-4656).
- Kernel, Apple iOS TM (before 9.3.5), Memory corruption leads to silently jail-break the device and install surveillance software.
- CVE-2016-4657 (https://web.nvd.nist.gov/view/vuln/detail?vulnId=CVE-2016-4657).

Memory corruption to execute arbitrary code or cause a DoS in Apple iOS TM (before 9.3.5). The vulnerability is exploited in the Safari WebKit that allows the attacker to compromise the device when the user clicks on a link.

Initially, the malware get started through a phishing message users can receive in their phones.

Once the user clicks or activates the message, it opens a web browser which eventually activates the malware payload silently without any notice from the user. The spyware payload includes enabling the attacker to expose different user applications such as: Messaging applications, emails, WhatsApp, FaceTime, etc.

2. Shedun, Hummingbad, Hummer, Shuanet, ShiftyBug, Kemoge, GhostPush, Right_core, and Gooligan

All those names of malwares in Android platform are believed to be connected with each other.

Malware investigators showed also that the majority of the source code for those different malwares is similar. Early versions of this malware appeared in 2015. They may represent aliases for the same original malware with some slight variations in: signature, payload, access method, functionalities, etc. The malware infects devices through the installation of some applications that include the malware. The add-on adware is added to normal and popular applications that users often download and use (e.g., Twitter, WhatsUp, Facebook, Google, etc.).

The new packaged applications are then made for public download in third party application stores. One of the payloads for the malware is exposing Google accounts for the victim user.

The malware series have also the ability to tamper with Android accessibility services. KitKat and Jelly Bean are the most widely impacted Android systems by this type of malware.

Malware payload can be also creating a backdoor to allow attacker to access victim mobile remotely. This is achieved by "auto-rooting" to elevate attacker privilege to a "root" privilege through exploiting some of the popular Android exploits (e.g., ExynosAbuse, Memexploit, and Framaroot).

3. Dendroid: A malware that targeted also Android platform. It was discovered in 2014 as a malware to allow attackers to gain remote control to victim mobile phones. Google Bouncer anti-malware that is deployed in Google Play was unable to detect this malware. It is believed that this malware uses remote control methods similar to those of earlier Zeus and SpyEye malwares. The malware was developed to be packaged with other legitimate mobile applications. As a typical remote-control malware, this malware can perform several types of actions on remote victims' mobile phones.

4. ViperRAT: A recent Advanced Persistent Threat (APT) malware that targeted Israel defense force. APT refers to the list of persistent malware threats with advanced and stealthy natures of hacking or malware activities. The term is used to label different categories or types of malwares in different platforms, not only in mobile devices. An APT can also include different malwares in different platforms. ViperRAT main goal was to spy on mobile victims, steal data and monitor audio and video recordings, contacts, and photos. Some versions of ViperRAT are believed to be deployed to victim machines through Social engineering methods using fake social networking profiles. Payload of the malware is installed as a Viber application update or as a general software update.

5. Pretender Applications: There are many incidents where fake applications exist in the mobile market application stores. Those fake applications use the name, logo, etc. of popular legitimate applications that users often use or download. This is an old and classical trick for some malwares for an easy deployment method. Users who download software applications from unknown market stores may not notice any difference between original and fake application. Even if some slight differences exist, they may think its just a new or enhanced version of the version they are familiar with.

6. Triada: This is a backdoor malware that targets Android systems. It also employed IP address spoofing in loaded web pages.

7. Hiddad: Another backdoor malware that targets Android systems. Similar to many other Android malwares, it repackages legitimate applications, include itself and then deploy using the same names of the legitimate applications in third-party application stores.

8. RuMMS: An android-based malware family targeted user in Russia via SMS Phishing (Smishing). SMS phishing messages that contain malicious links are

sent to victims mobile phones. The main payload in the malware is related to identity theft and stealing privacy and finance related information.

9. Brain Test: This Android application was in Google Play store for a while masquerading as a legitimate application. The application, believed to be primary for advertisement payloads, is masquerading as an IQ testing application. Google has "Google Bouncer" anti malware scanning program to scan applications in Google Play store. However, the Bouncer fails to detect the malware in this program for some time. In an advanced case, once the malware was detected, it was introduced again with an obfuscation mechanism to distort the ability to detect this malware as the earlier version.

10. XcodeGhost: An iOS malware that infects Xcode compiler and Integrated Development Environment, (versions 6.1 and 6.4). Xcode is used by iOS to develop mobile applications. The malware was first seen in the Chinese website and search engine Baidu. It is believed that more than 50 mobile applications that were developed and deployed were infected by this malware.

Some examples of the popular applications that were infected include: WeChat, Didi Chuxing, Railway 12306, WinZip, and Tonghuashun. An author claimed creating this malware and posted its code openly in Github. For users it was hard to judge if this is a malware as it came through legitimate applications and store. Most payloads discovered related to this malware are spyware related activities such as spying on users' private information, users' credentials, etc.

Table 2.3 shows examples of recent cyber threat trends.

Cyber Security Vulnerability Analysis

Some of the most popular websites that track vulnerabilities in US include:

- National Vulnerability Database: https://nvd.nist.gov/
- Security vulnerability database: https://www.cvedetails.com/

Repositories of this type of data, such as CVE Details and the National Vulnerability Database (NVD), are used to help consumers find this data in order to help them weigh their options. The numbers and types of these vulnerabilities represent a fair indication of the level of vulnerability exhibited by a given product. It's important to consider that these types of repositories are not definitive. They could not possibly contain a listing for every vulnerability, but rather only known and fully documented

Table 2.3 Examples of recent, emerging or popular cyber threats or attacks	Phishing attacks	Malwares (e.g., worms, ransomwares, etc.)
	SQL Injection (SQLi)	Denial-of-Service (DoS) and flooding
	Cross-Site Scripting (XSS)	Password/identity theft
	Man-in-the-Middle (MiM)	Supply chain attacks
	Advanced persistent threats	IoT Botnets

vulnerabilities. Because of this, a high number of vulnerabilities could also be an indication, or a by-product, of the popularity of a given product.

Several open source tools exist that can be used for vulnerability analysis of the different systems and applications. Following are examples of links that list examples of such vulnerability analysis tools:

- OWASP list of vulnerability analysis tools: https://www.owasp.org/index.php/Category:Vulnerability_Scanning_Tools
- Kali vulnerability analysis tools: https://tools.kali.org/tools-listing
- https://sectools.org/

Cyber Resilience

Cyber resilience refers to the ability to continuously deliver business functions despite adverse cyber threats or events.

Technology evolves, so do the volume and vigor of cyber-attacks. Hence, basic security will not be able to help and protect the enterprise. Some of the important required quality attributes to ensure cyber resilience:

- Recoverability: Information systems and assets should have the ability to recover from errors and attacks. This is a different stage than the role of security controls, e.g., firewalls, IDS, etc. where the goal of those controls is to try to prevent or protect against attacks. Recover methods consider what actions to take once such attacks or threats occur and what to do to limit the impact of those attacks and ensure business functions go back to normal.
- Adaptability: Information systems and their users should be able to adapt to the continuously changing environments and be able to accommodate new and different types of threats. With the continuous evolution of technologies, new products and services arise and show with them new types of threats. Users should not respond by avoiding those new products and services but rather learn how to adapt.
- Durability: Information systems and all applications should always be up-to-date with software fixes and updates.

K0029: Knowledge of Organization's Local and Wide Area Network Connections
K0047: Knowledge of Information Technology (IT) Architectural Concepts and Frameworks

Enterprise information security architecture frameworks represent a subset of enterprise architecture frameworks. Examples of existing security architecture frameworks include: The Open Group Architecture Framework (TOGAF),

SABSA framework and methodology, Zachman Framework. The U.S. Department of Defense (DoD) Architecture Framework (DoDAF), Open Safety & Security Architecture Framework, MITA, Extended enterprise architecture framework (EA2F), etc. The following links (http://www.iso-architecture.org/ieee-1471/afs/frameworks-table.html, http://thebrowsery.us/security-architecture-framework.html), provide a comprehensive list of Enterprise information security architecture (EISA) frameworks.

Different security frameworks or architectures may have different views and interactions between the different security areas and components as well as processes and activities. However, most of them agree on the need to have the following major areas: Authentication, Authorization, Audit, Assurance, Asset Protection, Availability, Administration, and Risk Management.

Security personnel should also plan for and prepare security artifacts that their adopted framework is proposing. The list can be small to include major artifacts such as: Security policies and guidelines, data classification, protection and ownership policies, risk, vulnerability, and gap analysis.

As one example of an IT architecture, we will demonstrate NIST model. Developed late-1980s by the National Institute of Standards and Technology (NIST) and others developed an IT reference architecture [1, 2], Fig. 2.3.

- Business Architecture level: Organization units that are dealing with external organizations.
- Information architecture level: This level focuses on information contents, presentation forms, and formats.

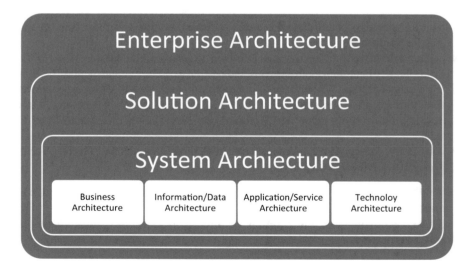

Fig. 2.3 An IT reference architecture example

- Information systems architecture level.
- Data Architecture level.
- Data Delivery Systems level.

K0049: Knowledge of Information Technology (IT) Security Principles and Methods (e.g., Firewalls, Demilitarized Zones, Encryption)

Security Principles

There are different types of security principles that are used to guide how to design/implement security controls and policies. We will cover examples of those in the next section.

- **Principle of Least Privilege**: This means that the default access is nothing for a user. Users will then be granted access to the resources that they need. In this regard, users are encouraged not to use administrator accounts all the time. If their accounts are exposed, with such high privileges, an attacker can significantly hurt the system. Alternatively, they should use normal or power user accounts and only elevate to administrators when they need to.
- **Separation of Duties**: Users should not accumulate responsibilities. They should not have open accounts that can play different roles in different occasions. Such roles should be divided.
- **Principle of Least Knowledge**: Similar to the principle of least privilege, users do not need to see resources that they have no associated tasks with. Intentionally or unintentionally users can abuse their privileges. In some phishing or social engineering types of attacks, those users can be victims and their accounts can be used without their consent knowledge to commit attacks on systems or expose their resources.
- **Defense in Layers**: Most security system designers consider protecting their systems using layers in different locations/artifacts within the systems or the enterprise. For examples, security and access controls can be found in routers, gateways, interior and exterior firewalls, operating systems, DBMS, anti-malwares, and applications. Those different security controls may see different aspects of the systems or networks, or different traffics and hence may take their actions (e.g., to permit or deny) interpedently. Access control decisions are approximately taken in all network levels or layers. Two main advantages can be observed in this non-centralized access control architecture:
- One access or security control needs not to be perfect and cover all system vulnerabilities and protect against all types of threats or malwares. This goal is not even realistic or practical.

- Different security or access control systems may take their decisions based on a local knowledge about an application, user, malware, etc. that may not be available to other access controls at the time of taking this promptly real time action.

The draw back of having such non-decentralized access control architecture is that decisions of the different access controls can be inconsistent and contradict with each other.

Most security controls implement control access to system assets and resources through Access Control Lists (ACLs). ACLs are mechanisms or concrete implementations of access control models.

ACLs represent permissions on system objects to decide who can have view/create/modify/execute a system resource or object. In operating system ACLs, an access control entry (ACE) is configured using four parameters:

- A security identifier (SID).
- An access mask.
- A flag about operations that can be performed on the object.
- A flag to determine inherited permissions of the object.

Firewalls

Firewalls are one of the most popular security controls in information systems. They are distinguished as lightweight actionable security controls that make decisions to block traffic based on a limited number of attributes in OSI Layers two and three (L2–L3 firewalls). However, this is the case in what can be referred to as "classical firewalls" as a new generation of firewalls operate on the application layer or layer 7 in OSI.

Rules in firewalls are generally classified under two main categories:

- Inbound rules: Rules that apply on incoming traffic. This is very important as our firewall main goal is to protect our system from harmful incoming traffic.
- Outbound rules: Rules that control traffic going out from our system or network.

While this category is less important, in general, that the first category for our system security, however, from a liability perspective, it is important to make sure that our system or network is not harming other systems or is used, without our consensus knowledge to attack others.

Web application firewalls. Those are also called Layer 7 or application layer firewalls. In comparison with classical L2–L3 firewalls, web application firewalls can look at different attributes that identify malicious applications, users, or traffic (Fig. 2.4).

The figure shows main attributes that should be specified in each firewall rule. Those include IP, MAC, and Port addresses for source and destination. If the user

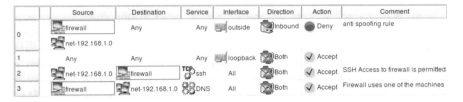

	Source	Destination	Service	Interface	Direction	Action	Comment
0	firewall net-192.168.1.0	Any	Any	outside	Inbound	Deny	anti spoofing rule
1	Any	Any	Any	loopback	Both	Accept	
2	net-192.168.1.0	firewall	TCP ssh	All	Both	Accept	SSH Access to firewall is permitted
3	firewall	net-192.168.1.0	DNS	All	Both	Accept	Firewall uses one of the machines

Fig. 2.4 Firewall rules, examples

selects the option (any, wild cards), then that will be a general flag with no specific source or destination. For example, the first firewall rule in indicates that all traffic going from the firewall (as an IP address) to any destination should be denied or blocked. Service option includes protocol using this traffic. Interface indicates the network card that the rule will be applied on.

The line below shows another example of a textual firewall rule. The line shows the same information in addition to specific inbound and outbound ports.

```
adapter A ip src addr xxx.xxx.x dst addr any tcp src
port 20 dst port 80
```

There can be usually some other options related to whether events should be recorded or not and some other optional features that may vary from one firewall vendor to another.

Firewalls and many other access controls use one of the four popular access control models: ABAC/OBAC, RBAC, MAC and MAC, each one of those have different implementations in the industry. For example:

- ABAC/OBAC: We described XACML as the most popular implementation of ABAC/OBAC.
- RBAC: Widely popular and has implementations in the different access control architectures such as Access Control Lists (ACLs) that can be found in firewalls, switches, routers, etc.
- discretionary access control (DAC): Different file systems use DAC to control users' access to the different files.
- Mandatory access control (MAC): Early implementations of MAC focused on multilevel security to protect military-oriented security classification levels with robust enforcement. Most of today's operating systems use DAC as their primary access control.

No access control model of those previously mentioned can be claimed as the best in all types of implementations. Certain quality attributes based on the usage of the model justify selecting one of the models in particular. For example, in many military applications two main criteria come first: confidentiality of certain information and greatly limit access to that information. For those two reasons hard-coded security in MAC is preferred. In many private companies, other quality attributes are more or as important: productivity, interorganizational.

data sharing, and information workflow between different. In such cases, DAC model can be a good choice. ABAC/OBAC recently evolves to take more market share especially due to its ability to limit or specific access based on more details on attributes related to (1) users who request the access, (2) accessed resources, (3) the environment or (4) the access context, etc.

K0050: Knowledge of Local Area and Wide Area Networking Principles and Concepts Including Bandwidth Management

Different computing systems (e.g., servers, desktops, laptops, smart phones, etc.) connect with each other using network devices such as switches and routers.

Switches are network hardware components that are used to connect local computers (i.e., Local Area Networks, LAN). Switches transfer or switch traffic from source to destination based on Layer 2 (i.e., MAC) information. As such, Access Control List (ACL) information in switches is accomplished based on MACs, Fig. 2.5, (Cisco Knowledge Base 2018).

VLAN ID attribute allows the creation of virtual LANs that redefine physical networks logically, rather than physically (i.e., based on the physical switches and topology). This concept is one of the key concept enablers in cloud computing and network virtualization.

Routers are very popular and key network components that connect different networks with each other. The whole Internet is about a large number of routers as intersections connecting the different computing machines, servers, networks, etc. with each other.

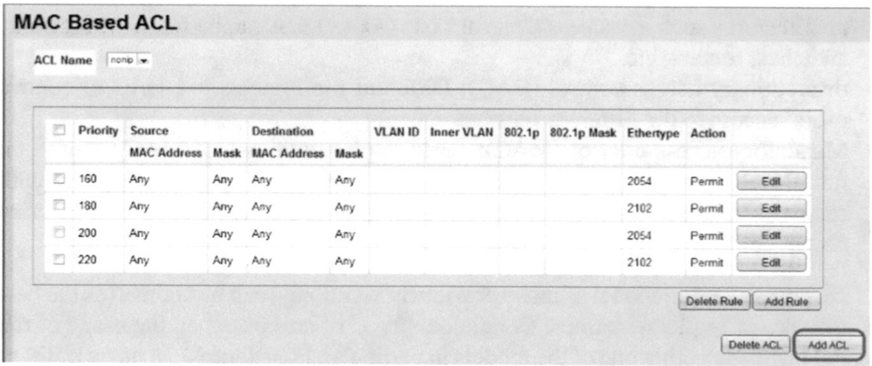

MAC Based ACL

ACL Name [nonb ▾]

	Priority	Source MAC Address	Mask	Destination MAC Address	Mask	VLAN ID	Inner VLAN	802.1p	802.1p Mask	Ethertype	Action
☐	160	Any	Any	Any	Any					2054	Permit	Edit
☐	180	Any	Any	Any	Any					2102	Permit	Edit
☐	200	Any	Any	Any	Any					2054	Permit	Edit
☐	220	Any	Any	Any	Any					2102	Permit	Edit

Delete Rule   Add Rule

Delete ACL   Add ACL

Fig. 2.5 MAC-based ACL, Cisco Knowledge Base 2018

```
v6acl#show ipv6 access-list
IPv6 access list ipv6acl
    permit tcp host 2001:AAAA::4 host 2001:BBBB::2 eq www sequence 10
    deny tcp any host 2001:BBBB::2 eq www sequence 20
    permit ipv6 any any sequence 30
v6acl#
```

Fig. 2.6 An ACL example in routers

While routers' main function is to transfer or route traffic from one location to another, deciding on the best route to do that, however they can also perform firewalling or access control functions. As such, many routers are sold with (built-in firewalls). Similar to firewalls they will have rules for inbound and outbound traffic. Similarly, they can filter traffic based inbound or outbound port numbers, IP addresses (V4 and V6) and MAC addresses (Fig. 2.6).

Network switches and routers include their own operating systems. Part of their operating systems, they also include access control systems or modules. In addition to access controls related to users, NOS may include access controls for other systems, applications, and even traffic. Host-based access control uses host IP, DNS, and possibly MAC addresses. Several network-based attacks related to those attributes are possible. Examples of those attacks include:

IP/MAC/ARP spoofing, DNS pharming, etc.

More recent software-controlled NOSs such as Software Defined Networking (SDN) Controllers aim at giving users and their applications a fine-grained level of access control on traffic from and to the network (For more details, see OpenDayLight project or platform at: https://www.opendaylight.org.

System and network administrators employ different tools to monitor traffic and bandwidth. Network and bandwidth management activities are necessary for security monitoring, polices and standards compliance, etc. Some of the tasks to be covered in bandwidth management:

- Quality of Service, (QoS): to monitor and ensure that network services are provided within acceptable ranges.
- Traffic shaping or rate limiting: To limit bandwidth, file sharing etc. for certain users, applications or in certain times.
- Scheduling algorithms.
- Traffic congestion avoid and optimization.
- WAN optimization: Ensure network **efficiency** at the large scale.
- Web caching: Improve network usage and Internet users' experience through cashing previously accessed web content.
- Track and monitor traffic and network usage for possible malware prediction (e.g., worms).

K0053: Knowledge of Measures or Indicators of System Performance and Availability

Systems and networks can be monitored for problems using different tools and applications. Those applications focus on a number of indicators or metrics that can show symptoms of network health or problems such as:

- Response time: Users interact with different systems or different system components. Response time to users when interacting requesting different services can be a very good indicator of system and network health.
- Throughput: In connection with response time, throughput measures the number of completed transactions per unit of time. In comparison with response time, throughput is goal or output oriented. For example, a throughput requirement may state that: System should be able to process more than 1000 transaction per second.
- Availability: Availability measures the overall percentage of time when systems, networks, services, etc. were available to users. For example, an availability requirement may state that: The cloud server should have an availability of 99.9% during business hours and weekdays.
- Reliability: Reliability is an important quality indicator that is related to continuity of operations without problems, failures, etc.
- CPU and Memory utilizations: Those are examples of specific indicators to the health of a computing machine (e.g., server, desktop, etc.).
- Bandwidth utilization: Bandwidth utilization is an indicator of network health and usage. It can be measured as an absolute or relative value. It can be used for benchmarking and historical monitoring of bandwidth usage or utilization over the weekdays, weekend, days, nights, etc.
- Failures rates: Similar to reliability, failure rates are indicators of networks or systems ability to provide services to users without problems or failures. Errors, failures, and faults are terms that can be used in this scope to indicate problems in networks or systems at different levels or details.
- Routers' up-time: This is an example of a quality indicator related to a specific network device.

K0094: Knowledge of the Capabilities and Functionality Associated with Content Creation Technologies (e.g., Wikis, Social Networking, Content Management Systems, Blogs)

In the second evolution of the Internet, users become not only information receivers but also creators. Online Social Networks (OSNs) such as: Google, Facebook, Twitter, Youtube, are now the most popular websites, Fig. 2.7.

Rank	Website	Category	Change	Avg. Visit Duration	Pages / Visit	Bounce Rate
1	G google.com	Computers Electronics and Technology > Search Engines	»	00:09:39	8.71	32.95%
2	youtube.com	Arts and Entertainment > TV Movies and Streaming	»	00:21:38	8.89	27.93%
3	facebook.com	Computers Electronics and Technology > Social Networks and Online Communities	»	00:11:00	10.63	31.27%
4	baidu.com	Computers Electronics and Technology > Search Engines	»	00:06:49	7.69	35.37%
5	instagram.com	Computers Electronics and Technology > Social Networks and Online Communities	»	00:06:24	13.81	35.98%
6	twitter.com	Computers Electronics and Technology > Social Networks and Online Communities	»	00:09:07	7.43	30.30%

Fig. 2.7 Most popular websites, 2019 according to similarweb.com

Normal users without any IT or technical knowledge can create pages in those websites and start creating contents (e.g., posts, pictures, videos, etc.). Those websites are open to all the public users around the world without limitations beyond age limitations.

The nature and domain of opportunities and businesses are expanding in those websites. Regular users can use those websites not only for social interactions but also for business interactions, marketing, etc.

The expansion and growth of content in those websites surpass content in classical media such as newspapers, magazines, radio stations, TVs, etc. for examples, some of the popular users or celebrities in OSNs can be more popular and have more audience that popular TV or movie stars and even popular than the TV stations themselves.

The impact of OSNs content on human lives and decisions make a serious concern with an ongoing investigation related to the possible usage of OSNs by Russia to manipulate US presidential elections in 2016. In one example, 3 Million Russian Troll Tweets are made available for public analysis and investigation (https://github.com/fivethirtyeight/russian-troll-tweets/).

References

1. Chief Information Officer Council. (2001). A practical guide to federal enterprise architecture version 1.0 preface. http://www.gao.gov/assets/590/588407.pdf, February 2001.
2. The Chief Information Officers Council. (1999). federal enterprise architecture framework version 1.1. http://www.enterprise-architecture.info/Images/Documents/Federal%20EA%20Framework.pdf, September 1999.

Chapter 3
IT Risk and Security Management

Izzat Alsmadi

K0002: Knowledge of Risk Management Processes (e.g., Methods for Assessing and Mitigating Risk)

In the field of information security, some of the terminologies or concepts used are misunderstood or in some cases are used synonymously. Some security concepts are closely related and worth being examined together. Widely related terms in information security include risk, threat, and vulnerability. In this section, we discuss about these concepts briefly.

- **Risks**: This is the likelihood that assets are being threatened by a particular attack. System-related security risks arise from the loss of confidentiality, integrity, or availability (CIA) of information or systems and reflect the potential adverse impacts to organizational operations, organizational assets.

  When it comes to risks, risk assessment is a critical step that considers the potential threats, the impact it causes, and the probability it occurs. Based on the risk assessment, the appropriate security measures can be considered. Table 3.1 shows a sample qualitative risk assessment model.

Table 3.1 A sample qualitative risk assessment model

Risk	Risk rate
>=K1	Very high
K2–K1	High
K3–K2	Medium high
K4–K3	Medium
K5–K4	Medium low
K6–K5	Low
K7–K6	Very low

© The Editor(s) (if applicable) and The Author(s), under exclusive license to Springer Nature Switzerland AG 2020
I. Alsmadi et al., *The NICE Cyber Security Framework*,
https://doi.org/10.1007/978-3-030-41987-5_3

- **Threat**: A security attack attempts to exploit vulnerability in the target, with the potential to adversely impact organizational operations, organizational assets, individuals, or other organizations via unauthorized access, destruction, disclosure, or modification of information. Security threats can take the form of worms, viruses, ransomwares, trojan horses, etc. To deal with threats, threat assessments focus on assessing threats and choose the best countermeasures as part of the organization wide security plan.
- **Vulnerability**: A vulnerability is a flaw in the design or configuration of systems or applications that can have security implications or consequences. Security breaches starts with an attacker searching/scanning targets for potential vulnerabilities, and then use such vulnerabilities to carry an attack. Testing for vulnerabilities is crucial for maintaining ongoing security assessment. A variety of organizations maintain publicly accessible databases of vulnerabilities (e.g., cvedetails.com, nvd.nist.gov, etc.
- **Assets**: An asset is a resource of value that is protection worthy. An asset can be tangible or intangible. Asset values can take different forms in addition to the monetary values. The value of assets decides the level or degree of risk mitigation.

The process of security and risk management and planning is extremely important when dealing with digital transactions. It is important for organizations to identify risks that may be included in future investigations. The approach for planning the risk activities is important and should be budgeted by the organization. Once the risks are identified, then the organization needs to establish a management process for the security concerns. Additionally, these concerns should be reviewed daily, weekly, or monthly depending on the organizational culture and risk tolerance. This is usually where the problem exists because risks are identified, but not managed with a consistent frequency. The approach should be to continually identify and control the risks. Security risks can be derived from web, mobile, network, database assets, and local activities.

Risk Management Approaches

- Mitigation
- Transference
- Acceptance
- Avoidance.

The decision to select an approach needs to be aligned with the overall organizational views and culture depending on the probability and impact of the risk. It is important to get the appropriate personnel involved if the risk can ultimately damage the organization's operations, reputation, or security.

Risk Tolerance

A key measurement for any organization is determining the overall risk tolerance. Some organizations take a proactive approach opposed to a reactive approach in regards to risk. A few things should be considered when deciding risk tolerance level is how long the information system has been in existence and what are the mitigating factors if potential issues arise. Organizations conduct risk assessments on the environment to determine the potential risk and probability of the occurrences. This takes a number of reviews by a team to outline and document the potential risks. Additionally, it needs an incremental approach to realize the potential events that can have negative outcomes.

K0048: Knowledge of Risk Management Framework (RMF) Requirements

Figure 3.1 shows main steps in NIST risk management framework (RMF), (NIST Special Publication 800-37, Guide for Applying the Risk Management Framework).

- **Categorize**: Determine the criticality of the asset according to potential adverse impact to the organization, mission/business functions, and the system.

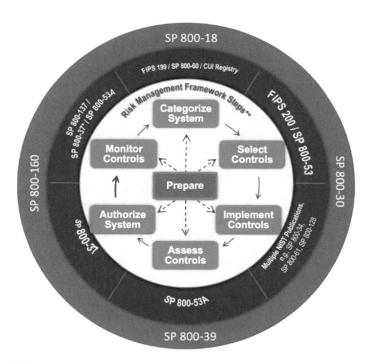

Fig. 3.1 NIST risk management framework (csrc.nist.gov)

Table 3.2 NIST SP 800-53 security controls

ID	Family	ID	Family
AC	Access Control	MP	Media Protection
AT	Awareness and Training	PE	Physical and Environmental Protection
AU	Audit and Accountability	PL	Planning
CA	Secunty Assessment and Authorization	PS	Personnel Security
CM	Configuration Management	RA	Risk Assessment
CP	Contingency Planning	SA	System and Services Acquisition
IA	Identification and Authentication	SC	System and Communications Protection
IR	Incident Response	SI	System and Information Integrity
MA	Maintenance	PM	Program Management

- **Select**: Select security controls starting with the appropriate baseline using selection process from previous step. NIST identified 18 categories of security controls (NIST SP 800-53), Table 3.2.
- **Implement**: Implement security controls within assets using sound system security engineering practices.
- **Assess**: Determine security control effectiveness and whether they are operating as intended, and meeting the security requirements for the system and environment of operation.
- **Authorize**: Examines the output of the security controls assessment to determine whether or not the risk is acceptable.
- **Monitor**: Continuously monitor the controls that are implemented for the system and its environment of operation for possible changes, signs of attack, etc., that may affect controls, and reassess controls' effectiveness.

NIST published other risk supporting publications such as: NIST SP 800-39: Managing Information Security Risk—Organization, Mission, and Information System View, NIST Special Publication 800-30, Guide to Conducting Risk Assessments, NIST Special Publication 800-37, Guide for Applying the Risk Management Framework.

K0149: Knowledge of Organization's Risk Tolerance and/or Risk Management Approach

Risk tolerance (also called risk appetite or risk propensity) is defined as the level of risk or the degree of uncertainty that is acceptable to an organization or that the organization is willing to risk. It can be translated to the amount of data, systems or assets that can be risked to an acceptable level. Risk tolerance is defined at the organization level and is implemented as part of risk management plan.

Some references differentiate between risk tolerance and risk appetite. The difference between risk appetite and risk tolerance is that risk tolerance represents the maximum risk that a company can reasonably assume while risk appetite represents the amount of risk that the company actually assumes based on their own acceptable levels.

Risk tolerance varies among organizations based on several factors, such as: the relative risk sensitivity of risk managers the organization's mission, and the nature of its assets, resources, and the operational processes that they support.

The first important step in the process is to define risk tolerance. Typical risk management approaches involve the steps to: (1) define risks under different categories, financial, organizational, technical, etc. risks, (2) prioritize risks, (3) decide actions and (4) take actions to respond to significant risks and (5) keep monitoring risks and risk mitigation activities.

The organization risk mitigation activities involve actions to either (1) prevent risks from occurring or lower their probability of occurrence, (2) prevent the impacts of risks once they occur or lower the level of impact. However, there are some risks that neither one of the previous mitigation actions are possible or doable. For those risks in particular, left after risk mitigation actions, (also called risk residuals), risk tolerance actions take place. In other words, organization responses to risks can be divided between risk mitigation and risk tolerance activities.

The acceptable levels of risk residuals can vary from one organization to another or one situation to another which implies difference tolerance actions for the same risk residuals.

Followings are examples of factors that can help define an appropriate level of risk tolerance [1]:

- Percent of critical assets for which a cost of loss, damage or disclosure has been quantified
- Percent of incidents that result in damage, compromise or loss beyond established thresholds
- Annual percent of interrupted business processes that were restored within targeted timelines
- Percent of security incidents that exploited existing vulnerabilities with known solutions.

K0165: Knowledge of Risk/Threat Assessment

Risk Assessment

Risks are assessed using quantitative and qualitative methods. Risks are assessed to (1) estimate their seriousness or likelihood to happen, (2) estimate their impact or liability once such risks become reality, and (3) estimate the cost of intervention or how much it costs to stop those risks, reduce their likelihood of occurrence, or stop/reduce their impact.

Risks and threats can come from any source: natural, humans, technical, security, etc. Our focus in this book is on security, security threats grow rapidly in the last few years to cover a large spectrum of malicious activities, mechanisms, applications, and so on. Different models define different stages in risk assessment, nonetheless, the process involves the following major stages:

- Risk identification
- Risk analysis
- Risk evaluation
- Risk monitoring and communication.

US FIPS 199 standard identifies three levels of risks:

- Low: The loss of confidentiality, integrity, or availability (CIA) could be expected to have a limited adverse effect on organizational operations, assets, or individuals.
- Moderate: The loss of CIA could be expected to have a serious adverse effect on organizational operations, assets, or individuals.
- High: The loss of confidentiality, integrity, or availability could be expected to have a severe or catastrophic adverse effect on organizational operations, assets, or individuals.

Risk assessment matrices take into consideration two main risk attributes: impact and likelihood, Table 3.3.

Table 3.3 An example of risk assessment matrix

Risk Matrix		Likelihood					
		Remote	Highly unlikely	Unlikely	Possible	Likely	Almost certain
Impact	Catastrophic	6	12	18	24	30	36
	Critical	5	10	15	20	25	30
	Major	4	8	12	16	20	24
	Moderate	3	6	9	12	15	18
	Minor	2	4	5	8	10	12
	Insignificant	1	2	3	4	5	6

Risk assessments can help organizations (NIST Special Publication (SP) 800-30: Guide for Conducting Risk Assessments):

- Determine the most appropriate risk responses to ongoing cyber-attacks or threats.
- Guide investment strategies and decisions for the most effective cyber defenses.
- Maintain ongoing situational awareness with regard to assets' security state.

Threat Assessment

Threat assessment is a structured process used to evaluate or determine the credibility and seriousness of a potential threat or risk typically as a response to an actual or perceived threat or concerning behavior.

Similar to risk assessment, threat assessment includes four major stages:

- Threat identification
- Threat assessment
- Threat case management (e.g., threat handling and mitigation activities)
- Monitoring and follow-up assessment.

K0195: Knowledge of Data Classification Standards and Methodologies Based on Sensitivity and Other Risk Factors

Data classification has been used for years to help organizations safeguard sensitive data with appropriate levels of protection.

U.S. government uses a three-tier classification standard in Executive Order 135261 and based on potential impact to national security if it is disclosed.

1. Confidential: Unauthorized information disclosure could be expected to cause damage to national security.
2. Secret: Unauthorized information where disclosure reasonably could be expected to cause serious damage to national security.
3. Top Secret: Unauthorized information disclosure could be expected to cause exceptionally grave damage to national security.

From security and privacy perspectives, data can be classified under different categories. The goal from such classification is to decide the visibility and access levels on those data whether in storage, processing or in transit.

Different names in different standards are used to identify high-level classified data (e.g., sensitive, classified, confidential, etc.).

Highly sensitive data is intended for limited and specific use by a specific target of audience. Explicit authorization by the data owners or managers is required for access because of legal, contractual, privacy, or other constraints. Confidential data has a very high level of sensitivity.

The following list include examples of sensitive data

- People credit card numbers, social security numbers, etc.
- Government or military sensitive information.
- Bank and financial information.
- Account credentials, user names and passwords.
- Law enforcement and investigative records.

Public information represents information that can result in a little or no risk in case unauthorized disclosure occurs. In most cases, public information is available to anyone who needs access to it. Examples of public information include but not limited to public news, press releases, maps, directories, and research publications.

K0203: Knowledge of Security Models (e.g., Bell–LaPadula Model, Biba Integrity Model, Clark–Wilson Integrity Model)

Role Based Access Control (RBAC) and Object Based Access Control (OBAC) are two most popular access control architectures or models used to develop access control systems. In addition to OBAC and RBAC some information systems utilize other models. For example, Mandatory Access Controls: MAC (such as Biba and Bell–LaPadula) where control is centralized and managed by system owners. In contrary with the centrality nature of MAC, Discretionary Access Control (DAC) is un-centralized and users can manage access controls on resources they own. One example of DAC model is NTFS permissions on Windows OS. On NTFS each file and folder has an owner. The owner can use Access Control Lists (ACL) and decide which users or group of users have access to the file or folder. Most today's operating systems use DAC as their main access control model.

Permissions: Permissions are the roles in an access control system to decide system constraints. The three main components in any permission include:

- Entity: The user/role or system that is granted/denied the permission.
- Action: This can typically include: view/read/write/create/insert/modify, etc.
- Object: This is the information system resource that will be the action object or where action is going to be implemented or executed.

Those are the minimum three elements that access control permission should have. Based on the nature of the access control system, many other optional elements can be included.

Bell–LaPadula is an old formal model of security policies that describes a set of access control rules. By conforming to a set of rules, the model tries to inductively proves that the system is secure.

The Biba integrity model [2] was published at Mitre after Bell–LaPadula model. Biba chose the mathematical dual of Bell–LaPadula policy wherein there are several integrity levels, a relation between them, and two rules which, if properly implemented, have been mathematically proven to prevent information at any given integrity level from flowing to a higher integrity level. Typical integrity levels are "untrusted", "slightly trusted", "trusted", "very trusted", etc.

K0214: Knowledge of the Risk Management Framework Assessment Methodology

From the perspective of Risk Management Framework (RMF), there are two main risk assessment activities:

- Evaluating the control implementation against the security controls and assessing any residual risks.
- There are also risk assessment activities that often follow NIST SP 800-30 or equivalent process to look at the system implementation in the context of all identified attack vectors and system vulnerabilities to identify risks.

These risks are assessed by the system owner/user. Those risks that exceed an acceptable risk threshold are re-evaluated. Generally, this would result in additional security controls implementation that is intended to adequately mitigate the risks.

Cybersecurity Risk Assessment Framework

Security team should periodically conduct assessment for their systems and assets. Those security and risk assessments can take different forms between technical assessments (e.g., vulnerability and penetration testing) and can also be accomplished through ad hoc based assessments (e.g., evaluating organization policies, conduct surveys, etc.).

CMMI institute cyber maturity framework offers a self cyber security risk assessment that companies can use to evaluate their security states or profiles. Examples of questions to evaluate in their survey:

- Whether the organization has identified potential physical vulnerabilities that might lead to known risks.
- Whether the organization has identified potential local vulnerabilities that might lead to known risks.

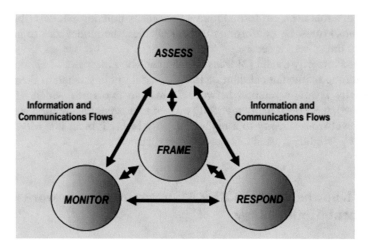

Fig. 3.2 NIST risk management process activities

- Whether the organization collaborates with relevant partners to periodically catalog known vulnerabilities.
- Whether a standard set of tools and/or methods exist to identify vulnerabilities.

NIST Special Publication (SP) 800-30: "Guide for conducting risk assessments" focuses on how to conduct risk assessments one of the four steps in the risk management process (Fig. 3.2).

NIST risk assessment divides activities in risk assessment into (Reader can refer to NIST Special Publication (SP) 800-30 for more details):

- Preparing for the risk assessment.
- Conducting the risk assessment.
- Maintaining the risk assessment.

Cyber security risk assessment framework is defined in [3] with six interdependent tasks:

1. Cybersecurity governance.
2. Inventory of information assets: data, infrastructure, applications.
3. Standard security configurations.
4. Information access management.
5. Prompt response and remediation.
6. Ongoing monitoring.

DITSCAP, DIACAP and RMF Certification and Accreditation (C&A)

DoD has their very first risk certification and accreditation (C&A) process: Defense Information Technology Security Certification and Accreditation Process (DITSCAP) in 1997. It requires a lot of documentations, and systems or assets were treated interpedently in the enterprise.

DIACAP replaced DITSCAP in 2007 which is more enterprise centric and has less documentation than DITSCAP. However, the standard was not adopted by other government organizations which cause interoperability issues.

In 2013 DoD Risk Management Framework (RMF) replaced DIACAP, (Table 3.4). RMF changed also the process name from Certification & Accreditation (C&A) to Assessment & Authorization (A&A). NIST released a standard related to RMF: NIST SP 800-37 Revision 1.

NIST lists specific roles involved in the A&A process with mapping to DIACAP processes, such as:

- The Risk Executive Function (maps to DIACAP's PAA)
- Authorizing Official (maps to DIACAP's DAA)
- Security Control Assessor (maps to DIACAP's CA)
- Information System Security Officer (maps to DIACAP's IAM).

K0232: Knowledge of Critical Protocols (e.g., IPSEC, AES, GRE, IKE)

Network encryption at the network or data link layer is not visible to users and their applications. Internet Protocol Security (IPSec), a set of open Internet Engineering Task Force (IETF) standards is an example of encryption at this level. The goal is to create private communication over IP network. Encryption can occur at the packet level.

Virtual Private Networks VPNs are created through encrypting the traffic using virtual tunneling protocols. This enables remote connection or remote users to connect using the public Internet. Their traffic will be tunneled and encrypted independent

Table 3.4 DIACAP versus RMF

DIACAP	RMF
Information assurance	Cyber security
Classification levels C&A	Assessment and authorization
DAA decision making	AOs decision making
Interim Authority to Operate (IATO)	No IATO

from the rest of Internet traffic. An attacker who is trying to sniff or spy on the data will only see encrypted data which will ensure its confidentiality. This can also prevent unauthorized users from using the VPN.

VPN uses IPSec as a secure protocol for IPv6. TLS/SSL is also used to encrypt network traffic in some applications such as OpenVPN. Cisco uses another protocol: Datagram Transport Layer Security (DTLS) for UDP traffic. Secure Socket Tunneling Protocol (SSTP) is used by Microsoft for Point to Point Protocol PPP or communication.

Advanced Encryption Standard (AES) is a recent cryptographic algorithm to protect sensitive information. AES is a privacy transform for IPsec and IKE that was developed to replace former DES and be more secure and harder to break. AES has a larger key size, while ensuring that the only current known approach to decrypt a message is for an intruder to try every possible key. AES has a variable key length—the algorithm can specify a 128, 192 or 256-bit key.

Generic Routing Encapsulation (GRE) is a protocol that encapsulates packets in order to route other protocols over IP networks. GRE is defined by RFC 2784. GRE tunnels sometimes are combined with IPSec, because IPSec does not support IPv6 multicast packets. This function prevents dynamic routing protocols from running successfully over an IPSec VPN network. The IPv6 over IPv4 GRE Tunnel Protection feature allows IPv6 multicast traffic to pass through a protected generic routing encapsulation (GRE) tunnel.

Internet Key Exchange (IKE) is a key management protocol used to authenticate IPsec, negotiate, and distribute IPsec encryption keys, to automatically establish IPsec security associations (SAs).

IKE negotiation comprises two phases: Phase 1 negotiates a security association between two IKE peers, which enables the peers to communicate securely in Phase 2. During Phase 2 negotiation, IKE establishes SAs for other applications, such as IPsec.

K0263: Knowledge of Information Technology (IT) Risk Management Policies, Requirements, and Procedures

Developing an IT risk-management policy provides organizations with the security to handle customers' sensitive data. Managing IT risks is a structured process that involves a series of activities that are designed to: identify, assess and mitigate risks, develop response plans, and review risk management procedures.

The followings are activities to consider when developing such policies:

- Catalog IT related organization's assets.
- Determine possible threats on assets.
- Estimate the cost of threats assessment and mitigation.
- Prioritize risks and for those that need action, determine security controls which could mitigate each one of the selected risks.

- Estimate the cost of each security control implementation.
- Compare the costs of each risk and its corresponding security control.
- Implement the risk controls that are cost-effective (e.g., using risk probability of occurrence, level of impact, and cost of intervention).
- Educate users on the new security controls, policies, and procedures.
- Create a process to track how risk-management controls are being implemented and how vulnerabilities are addressed.

IT policies and procedures describe the importance of managing IT risks. They can be part of risk management and business continuity plans.

Security policies and procedures can also assist staff training on issues such as

- Phishing and spam emails, safe email use.
- Setting out processes for common tasks, communication, access control, etc.
- Controlling and managing changes to IT systems and assets.
- Responses to IT and security incidents.

IT policies can also provide users with proper assets' usage behaviors and define acceptable behaviors in relation to key IT issues, services, assets. Security controls, such as systems that limit access to sensitive data or installation of software, are also important elements in IT security policies.

K0281: Knowledge of Information Technology (IT) Service Catalogs

IT service catalogs are databases or structured documents with information about all Live IT Services (including those that are available for deployment). IT service catalogs are established tools for IT service managers. The most important benefit of using those catalogs is the documentation of functions and quality. A broad range of different approaches for structuring service catalogs exist. Each one includes different information to cover for each IT service or application. Table 3.5 shows examples of IT service catalogs [4].

K0295: Knowledge of Confidentiality, Integrity, and Availability Principles

The CIA-triad—confidentiality, integrity, and availability—is no longer considered as an adequate and complete set of security goals. The CIA-triad does not cover new threats that emerge in the collaborative de-parameterized environment [5]. A new list of security goals is discussed in this section that accounts for all current information security threats.

Table 3.5 Examples of IT service catalogs [4]

Service type	Service	Service description	Service request	Service request description	Costs
Corporate	Email	Corporate email is the primary mechanism for facilitating communication throughout the organization	Standard Email Account	The standard email account includes all of the features defined above plus: • 512 MB of storage per user • Optional Blackberry support using Blackberry Enterprise Server	Client maintenance server (HW) server (SW) maintenance (HW) maintenance (SW) storage LAN WAN
Corporate	Service desk	The service desk is designed to be the single point of contact for users to request additional services or report issues with existing services to IT	Report service degradation or outage (Incident)	The service desk will record, classify, prioritize, and resolve if possible, any Incidents reported by users	Physical space phones ACD/IVR desktop/laptop service desk SW server (HW) maintenance (SW) maintenance (HW) staff
Corporate	Telephony	This service provides the basic telephone functionality, plus some others functional attributes like voice mail, call transfer, conference call, two-line entry, intercom, manager-line status, and related components	Fulfillment/provisioning of equipment	The mechanism by which users can request telephony services	Telephone switch ACD/IVR

- **Confidentiality**: in the context of information security, confidentiality means that information should be secured and only authorized users may access and read such information. Confidentiality is very important to individuals as well as organization since violating this goal can lead to devastating consequences. Security threats to information confidentiality are malware, social engineering, network breaches. Since confidentiality is really important, mechanism for protecting information should be in place. Security measures for maintaining information confidentiality include but not limited to cryptographic techniques and access controls.
- **Integrity**: integrity is all about making sure that the information has not been modified or corrupted. Data integrity covers data in storage, during processing,

and while in transit [6]. Information integrity involves ensuring information non-repudiation and authenticity. In this context, source integrity plays an integral role in validating information integrity. Source integrity is defined as making sure that the sender of the information is who it is supposed to be [7]. Spoofing is one of known threats to integrity where the attacker deceives the receiver and supplies incorrect information. Information integrity protection mechanisms are grouped into preventative mechanisms and detective mechanisms. Example techniques for protecting information integrity include digital signatures and hash algorithms.

- **Availability**: availability is all about ensuring timely and reliable access to and use of information [6]. The integrity and conditionality of information is worthless if such information is not available for the intended users. Threats to information availability is not only technical one such as Denial of Service (DoS) attack, they also include natural and manmade disasters such as tornados and fire. One of the key security measures to protect against threats to information availability is backups.
- **Identification**: Identification is the first step in the identify-authenticate-authorize sequence that is carried daily whenever a user is accessing information. The identification all about calming that you are somebody. For example, one can claim that he is John Doe. In the context information security, identification is similar to providing a username or something else that uniquely identifies a user. Nowadays, the most commonly used identification method is user ID.
- **Authentication**: authentication is the second step in the identify-authenticate-authorize sequence that verifies the authenticity of the claimed identity. During authentication, the user proves that you are indeed the one you claim to be. Authentication occurs often as a prerequisite to allowing access to resources in an information system [8]. In case of John Doe, the ID card is used to authenticate the claim he is John Doe not someone else. In the context of information security, there is different ways to authenticate users such as something you know (passwords), something you have (smart card), or something you are (biometrics). In certain situations, and depending on the sensitivity of the information, a combination of these authentication methods can be used.
- **Authorization**: access privileges granted to a user, program, or process or the act of granting those privileges [9]. Once the user is identified and authorized, they are assigned a set of permission and privileges—known as authorization—that defines what they can do with the information and the system. Unlike identification, which requires some sort of username, and authentication, which requires for example password, authorization is implemented part of the security policy within the organization. The most commonly used approach for granting permissions once users are authorized is role-based access control in which permissions are associated with roles, and users are made members of appropriate roles [10].
- **Accountability**: the security goal that generates the requirement for actions of an entity to be traced uniquely to that entity. This supports non-repudiation, deterrence, fault isolation, intrusion detection and prevention, and after-action recovery and legal action [6].

- **Privacy**: privacy is the process of restricting access to subscriber or relying party information in accordance with federal law and agency policy [11]. In the context of information security, privacy is critical in case the information identifies human beings. Such information includes social security number, name, and address. Information security policies play a major role in protecting personal information, and how such information is collected and processed.
- **Non-repudiation**: is the security service by which the entities involved in a communication cannot deny having participated. Specifically, the sending entity cannot deny having sent a message, and the receiving entity cannot deny having received a message [12].

K0297: Knowledge of Countermeasure Design for Identified Security Risks
K0298: Knowledge of Countermeasures for Identified Security Risks

Security countermeasures must be implemented to prevent threat agents from successfully achieving their goals. Major threat agent goals include the achievement of:

- Unauthorized access.
- Unauthorized modification or destruction of important information.
- Denial of service or authorized access.

If security countermeasures are not sufficient to prevent the threats, the existence of the countermeasures is not effective and useful. Hence it is important to design relevant security measures capable of preventing targeted threats or attacks. Countermeasures can be physical devices, administrative procedures etc. that may become ineffective.

It is important to consider not only the capability of countermeasures to address threats or vulnerabilities, but also any potential side effects on system operation as a result of the security control design. Invasive countermeasures can effectively address vulnerabilities, but may also unacceptably impact normal system operation.

The National Institute of Standards and Technology (NIST) has identified set of categories of Recommended Security Controls to protect and support the appropriate security level of information. The controls are safeguarding measures to protect the critical and vital operations of information Systems; make it dependable; more resistant to attacks; limit the damage when attacks occur; and make the systems more robust. Federal Information Processing Standard Publication (FIPS 199) has established three categories (High, Moderate, Low) of a potential impact on an organization and individual if the information system is exposed. The problem with this matrix is that there are known risk and unknown risk, with known risk as Knight stated, it is "easily converted into an effective certainty," while unknown risk or uncertainty is "not susceptible to measurement.". Many people deal with risk as—conditions, but according to the PMBoK, 2000 edition, risk is "an uncertain event or condition that, if it occurs, has a + or − effect on a project objective.".

The Risk matrices (likelihood and impact) are used for risk management decisions in many organizations. They have instinctive appeal to be used and understood, but risk matrices that are based likelihood of occurrence and their impacts have limitations associated with subjective assessments related to subject matter experts (SME). The value of SME is in helping the development of models representing the interrelations between variables to reflect valuable representation of risk management information.

K0326: Knowledge of Demilitarized Zones

Demilitarized zones are assets or network segments located as perimeter defense points and are used to isolate internal protected network elements from external, untrusted sources. Network designers should use demilitarized zones to house systems and information directly accessed externally. In other words, if certain information and systems are required to be accessed from the Internet or some other form of remote access, they should be placed in the less trusted demilitarized zone.

DMZ adds an extra layer of security to an organization. A monitored network node that is externally visible/reachable can access what is exposed in the DMZ, while the rest of the organization's network is safe behind the firewall or other security controls. Hosts in the DMZ have tight controlled access permissions to other services within the internal network, because data passed through the DMZ is not very secure.

All services accessible to users communicating from an external network should be placed in the DMZ. The most common services are:

- Web servers
- Mail servers
- FTP servers.

K0383: Knowledge of Collection Capabilities, Accesses, Performance Specifications, and Constraints Utilized to Satisfy Collection Plan

Full spectrum refers to the overall understanding (e.g., intelligence collection capabilities) and countering of threats across the Cyber Electronic Warfare (EW) domain or spectrum with the goal of providing full cyber control of allies and denying those to adversaries (i.e., both defensive and offensive cyber activities). It refers to including research, knowledge and development of sensors, concepts, techniques and technologies encompassing collection, exploitation and engagement of all data and signals across the Cyber EW spectrum. Full spectrum cyber is a term coined by the DoD to include both defensive and offensive cyber operations. Full spectrum cyber refers also to the full cyber support life-cycle: from providing network and systems design to operational support, security intelligence and cyber training, and exercise support.

In US, one military unit, U.S. Army Cyber Command (ARCYBER) provides cyber soldiers to support military missions. These Soldiers are tasked with defending army networks and providing full-spectrum cyber capabilities.

CNA/D/E/O

Full spectrum capabilities try to integrate elements from: Computer network defense (CND) with offense: attack and exploitation (CNA/E) into one platform.

- Computer network attack (CNA) indicate actions taken through the use of computer networks to disrupt, deny, degrade, or destroy information resident in computers and computer networks, or the computers and networks themselves (DoD-JP-2010).
- Computer network defense (CND). Actions that are taken to protect, monitor, analyze, detect, and respond to unauthorized activities within information systems and computer networks, (DoD-JP-2010).
- Computer network exploitation (CNE). Enabling operations and intelligence collection capabilities that are conducted through the use of computer networks to gather data from target or adversary information systems or networks, (DoD-JP-2010).

Full spectrum analysis requires multi-INT analysis approach. Multi-INT (i.e., multiple-intelligence) is the fusion, integration, and correlation of different types of data collected from different sources to provide a full operating view. The main two intelligence components to integrate are SIGINT, GEOINT, and MASINT. Open source and social media data are also important recent components. More recent components evolved such as: activity-based intelligence (ABI).

- Computer network operations (CNO)

In addition to multi-INT in terms of the different sources or methods of collecting intelligence data, multi-INT should employ sharing and operations:

- Cross-agency multi-INT sharing: Between the different intelligence agencies, public and private sectors. One example of such efforts a project called MISP—Open Source Threat Intelligence Platform & Open Standards for Threat Information Sharing, www.misp-project.org
- Cross-domain multi-INT operations: Ideally, this should be in the form of autonomous or self-adaptive security controls that learn threats in the domain and adapt itself to counter such threats.

K0388: Knowledge of Collection Searching/Analyzing Techniques and Tools for Chat/Buddy List, Emerging Technologies, VOIP, Media Over IP, VPN, VSAT/Wireless, Web Mail, and Cookies

Sources of Cyber Intelligence or Collection Capabilities

Currently, there are several categories or sources of cyber intelligence, also called collection capabilities, or intelligence gathering disciplines. This list continuously grows vertically and horizontally.

- *Open source intelligence (OSINT)*: Cyber intelligent team should learn how to gather data points, transform these data points into actionable intelligence that can prevent target attacks. They should learn how to identify, repel, or neutralize targeted intelligence gathering against organizational assets. OSINT includes data collected from publicly available sources, free or subscription-based, online or offline.

  OSINT can include many sub-categories such as:

- **Classical media**: Such as newspapers, magazines, radio, and television channels.
- **Online Social Networks (OSN) or Social media intelligence (SOCMINT)**: Blogs, discussion groups, Facebook, Twitter, YouTube, etc.
- Internet public websites and sources.
- Communication Intelligence (COMINT).
- Measurement and signature intelligence (MASINT).
- Search engines (e.g., Google, Yahoo, etc.).
- **Deep or dark web intelligence**.

Deep web: Those include web-pages, documents, etc. that are not indexed by main search engines and/or that cannot be read or accessed by conventional methods.

In percentage, the public or visible web is much smaller than deep web. Deep web can include the following categories: Dynamic web pages, Blocked sites, Unlinked sites, Private sites, Non-HTML or Scripted content, and Limited or local access networks or content not publicly accessible through the Internet.

Dark web or net: Those include web-pages, documents, etc. that are accessed by anonymized methods (e.g., TOR browsers) and are often used for criminal activities.

The dark web has become a port for hacking communities, offering cyber criminals the ability to discuss offer and sell new and emerging exploits (e.g., zero-day vulnerabilities or exploits). 0 day forum is a popular example of darknet websites (website link continuously varies, e.g.: http://qzbkwswfv5k2oj5d.onion.link/, http://msydqstlz2kzerdg.onion/, etc.).

Some of dark web forums are accessible only via the TOR network, while others are accessible via traditional web browsing. Those dark web forums start to have their own strict vetting processes to ensure that they will not be targeted by intelligent teams and face criminal charges and legal consequences. As such, it is common to have some users in those websites who are decoy intelligent personnel, police officers, FBI, etc.

The website is a market for buying and selling zero-day vulnerabilities. In addition to zero-day vulnerabilities, these forums offer a variety of "services" ranging from illegal drug sales, forged items (e.g., passports, driver licenses, credit cards, bank notes), weapons, identity theft information (e.g., PII; personal identifiable information), or botnet services.

For security intelligence, one of the main goals to study dark webs is to develop a functioning system for extracting information from those communities and apply machine learning methods to predict cases of considerable threats. The fact that humans heavily depend and use the Internet these days in all life aspects, gives hackers a platform rich of data and resources for hackers to collect data and learn how to hack and attack users and information systems. Not even dark web sites, but public websites can be also used as effective hacking or attacking tools. For example, websites such as: Shodan: (https://www.shodan.io/), Zomeye: (https://www.zoomeye.org), and https://www.go4expert.com/ can provide a wealth of information for attackers about candidate targets with very good introduction details to start further investigations and analysis.

NSA XKeyscore Program

This is a web-based tool that can collect public and private information (e.g., emails, chats, etc.) without the need for authentications. Majority of details on this program is not disclosed but claim the ability to collect a large amount of private information about persons of interests (including email contents, private chats, Internet search terms, Internet browsing history and activities, etc.). Under FISA act, NSA is required to request a warrant if the person of interest is a US citizen.

K0446: Knowledge of How Modern Wireless Communications Systems Impact Cyber Operations

Currently, mobile phones provide services beyond the classical phone calls. Those services are gradually converging to the same services can be offered by computing desktops or laptops. On the other hand, as smart phones accompany users almost everywhere, they can provide valuable location-based information.

Wireless transmissions are not always encrypted. Information such as e-mails sent by a mobile device is usually not encrypted while in transit. In addition, many applications do not encrypt the data they transmit and receive over the network, making it easy for the data to be intercepted. For example, if an application is transmitting data over an unencrypted WIFI network using http (rather than secure http), the data can be easily intercepted. When a wireless transmission is not encrypted, data can be easily intercepted by eavesdroppers, who may gain unauthorized access to sensitive information (e.g., host computers) without the need to be host administrators, power users or even users in those local hosts. Their root level role implies having an administrator privilege in all network or system resources.

You could connect to an unsecured network, and the data you send, including sensitive information such as passwords and account numbers, could potentially be intercepted. Many attackers can possibly create "free WIFI" networks to be used as honey-bots. They can provide users with free internet access while intercepting and spying on their sensitive data. In mobile operating systems it is unconventional to allow users to have "root" access to the operating system. This "privilege elevation to root" is called Jail-breaking. Users are not recommended to try to perform Jail-breaking, using certain methods or tools. This may open the operating system for several possible vulnerabilities and break the pre-designed operating system security architecture [13].

A user of an Apple mobile phone can have the ability to wipe their phone remotely using a mobile device management (MDM) server and iCloud or ActiveSync. All files will be then inaccessible and new user may need to create a new encryption with their new OS installation. System can be also automatically wiped after several unsuccessful PIN attempts. They can also lock their phones. In recent investigations, there were some cases that FBI requested access to credentials of iPhone that were locked (e.g., see [14]). On the other hand, black market provides all types of support and mechanisms to jail-break phones and counter activation or service provider locks.

Using smart phones is convenient and no one can afford abandoning them these days. On the other hand, they can be a weak point through which hackers or adversaries can target or track humans and their data. Their weakness does not only arise from the fact that they can be always used to track human targets, but also because they tend to have weaker overall security controls when compared with desktops or laptops [15].

Smart phones most popular attack types include: scams, phishing, data stealing and spying apps, malware, and ransomware.

K0506: Knowledge of Organization Objectives, Leadership Priorities, and Decision-Making Risks

Risk involves the possibility that a decision will result in a negative outcome; taking a chance; risk may involve losing money. All decisions involve consequences and many involve risks. In the decision-making scope, risks are potential obstacles that may prevent you from getting what you want. For example, you may make less beneficial decisions as they involve less risks that other alternatives. Nonetheless, we should not make risks as the main factor in our decision making and ignore the need to balance this with main business goals or functions.

Decision-making is very challenging in the current business environments due to several factors such as: Competition, growing demands, supply and acquisition, outsourcing of functions, and decentralization of decisions. These external and internal factors may directly complicate companies' decisions-making processes that have to find the right balance between business agility and the necessary controls over financial and non-financial actions exposing organizations to decision-making risks.

A risk-based decision involves making a decision with a known risk. For example, a company made a decision not to roll out patches this cycle because there were no known exploits in the wild. However, you got hit by a 0 day exploit never exploited before.

K0527: Knowledge of Risk Management and Mitigation Strategies

One main stage in risk management is risk mitigation. Risk mitigation involves taking actions for serious risks to:

- Stop the risk from occurrence or reduce the occurrence likelihood.
- Stop the impact of the risk once it occurs or reduce the level of impact.

There are different approaches to risk mitigation. Here are the most common methods:

- Assume and/or Accept: Security personnel may identify and acknowledge certain risks and decide to allow the potential threat to follow through without deciding specific strategies to manage the situation. Although no action is involved, this action still represents a considered decision.
- Watch and Monitor: After the initial process of identifying the potential risks, decisions can be made to keep some risks on the radar. There are different reasons for such decisions. In general, all risks whether we take actions against or not should be watched and monitored. However, risks in this category will be watched and monitored without any further actions. They can be serious risks ranked high but just showed up. So, they may be possible transient risks. In other situations, some risks may rank low in early cycle then move up gradually, or the opposite. For all those reasons, whether we want to take actions to mitigate risks or not, all serious risks should be continuously monitored.
- Avoid or limit risks: In this choice, actions are taken to prevent such risk from occurring. Avoiding risks completely may not be always a viable choice.

Alternatively, our actions may lower the probability or likelihood of risk occurrence.

- Warn or buffer: In some situations, action to take on certain risks is to provide early warning once such risk is approaching. This is very popular in natural disasters (e.g., flooding, hurricanes, earthquakes, etc.).
- Risk control that involves measures and actions to block and stop the threat or security risk. A good example of such measures is a firewall that will block malicious traffic and malwares.
- Risk transfer: In some cases, risks can be unavoidable, decisions can be made to provide methods to reduce the risk or threat impact.

Which action to take for a particular threat or risk? It depends on overall risk assessment or management activities that include risk identification, likelihood, level of impact, and cost of intervention. It depends also on the value and nature of the assets that can be impacted by those risks. Some security controls can serve the whole organization or enterprise and can protect against a large inventory or risks. Some serious assets with sensitive information should be protected using different risk mitigation strategies (e.g., avoidance, control and transfer, etc.).

References

1. Campbell, G. (2013). Metrics for success: Aligning incident impact with "Acceptable" risk, what is your organization's risk tolerance? Security info. Watch (www.securityinfowatch.com), June 14, 2013.
2. Biba, K. J. (1977). Integrity considerations for secure computer systems, USAF Electronic Systems Division.
3. The Institute of Internal Auditors. (2016). guidance@theiia.org, Assessing cybersecurity risk roles of the three lines of defense.
4. Emtec Boot Camp Web Event, Crawl, walk, run, approach, IT service catalogue, May 2011.
5. Cherdantseva, Y., & Hilton, J. (2013). A reference model of information assurance & security. In *2013 Eighth International Conference on Paper Presented at the Availability, Reliability and Security (ares)*.
6. Stoneburner, G., Hayden, C., & Feringa, A. (2001). Engineering principles for information technology security (a baseline for achieving security).
7. Matteucci, I. (2008). *Synthesis of secure systems*. PhD thesis, University of Siena (April 2008).
8. NIST. (2013). Security and privacy controls for federal information systems and organizations: National Institute of Standards and Technology (NIST).
9. Dufel, M., Subramanium, V., & Chowdhury, M. (2014). Delivery of authentication information to a RESTful service using token validation scheme: Google Patents.
10. Sandhu, R. S., Coyne, E. J., Feinstein, H. L., & Youman, C. E. (1996). Role-based access control models. *Computer, 29*(2), 38–47.
11. Kuhn, D. R., Hu, V. C., Polk, W. T., & Chang, S.-J. (2001). Introduction to public key technology and the federal PKI infrastructure.
12. NIST. (1994). Federal Information Processing Standard (FIPS) 191: National Institute of Standards and Technology (NIST).
13. Alsmadi, I., Burdwell R., Aleroud A., Wahbeh A., Al-Qudah, M. A., & Al-Omari, A. (2018). Practical information security. Cham: Springer.

14. CBS News, April 21, 2016. http://www.cbsnews.com/news/fbi-paid-more-than-1-million-for-san-bernardino-iphone-hack-james-comey/.
15. Kimberly Underwood, DHS Builds Mobile Defenses, the cyber edge, July 1, 2018.

Chapter 4
Criminal Law

Chuck Easttom

Introduction

While it is certainly true that computer security goes well beyond merely obeying the law, knowledge of relevant laws is important. Compliance with the law is a critical part of information security. Criminal law intersects with cyber security in the areas of digital forensic investigations and incident response. It is important for cyber security professionals to be conversant in these laws. In this chapter we will explore the major laws in the United States, as well as briefly look at some individual state laws. There will also be a brief mention of those international laws that frequently impact cybersecurity. The focus will be on criminal law, not civil law. With some of the laws discussed in this chapter, one or more example cases are briefly discussed to help clarify the laws intent and use.

General Cybercrime Laws

Later in this chapter we will review some laws that address specific types of cyber-crime. However, there are some broadly worded federal statutes that can be applied to a wide range of computer crimes. We will look at these laws first. These are among the most important for those investigating incident response as they are typically used to prosecute hacking related offenses.

Computer Fraud and Abuse Act (CFAA) 18 US Code §1030

In the United States, this is a fundamental computer crime law. This law is was passed in 1984 and is frequently referred to as the Computer Fraud and Abuse Act.

© The Editor(s) (if applicable) and The Author(s), under exclusive license to Springer
Nature Switzerland AG 2020
I. Alsmadi et al., *The NICE Cyber Security Framework*,
https://doi.org/10.1007/978-3-030-41987-5_4

While the law has more nuance and specificity, the essence of it was that it made accessing a computer without authorization or in excess of authorization a crime. The law was specifically covering computer systems that had some federal nexus. Such as computers that were owned or controlled by some federal agency, or those used by financial institutions. This law has been updated numerous times including in 1994, 1996, 2001, and in 2008.

Section 1 of the law prohibits obtaining specific information from a computer without authorization. The information covered includes financial data, any government data, and the vaguely worded "information from any protected computer". Section 5 states "knowingly causes the transmission of a program, information, code, or command, and as a result of such conduct, intentionally causes damage without authorization, to a protected computer;" This covers malware creation. Section 6 addresses trafficking in passwords or similar information. Section 7 is rather interesting as it addresses computer blackmail and extortion.

The amendments to the act made in 2008 eliminated the requirement that the information stolen must involve interstate of foreign communication. Also, the 2008 amendment eliminated the requirement that loss must exceed $5000.

In 2008 this law as used in the case of United States v. Drew. In 2006 Lori Drew reportedly became concerned that a teenager named Megan Meier, who know Mrs. Drews daughter Sarah, had been spreading false information about Sarah Drew. Lori Drew then allegedly created a Myspace account for a non-existent 16-year-old boy to contact Megan and to mislead her. On October 16, 2016 the fake account for the non-existent 16-year-old boy sent a message to Megan effectively encouraging suicide. Megan subsequently committed suicide. In December 2007 Missouri state prosecutors declined to prosecuted Lori Drew in connection with Megan's death. The US Attorney used the Computer Fraud and Abuse Act to charge Mrs. Drew. She was eventually acquitted.

More recently, in 2019 the Computer Fraud and Abuse Act is being used against Julian Assange. Specifically, he is being charged with conspiracy to violate the Computer Fraud and Abuse Act. This is due to his publication of stolen information. Whomever actually stole the information would presumably be guilty of violating the Computer Fraud and Abuse Act.

These two cases are quite different, with substantial differences in the underlying acts, and the entire case. These illustrate the broad nature of the Computer Fraud and Abuse Act. It applies to a wide range of computer crimes.

When performing incident response in any organization, you will often wish to evaluate whether or not the incident involves a violation of 18 US Code 1030. It is not expected that you are an attorney and should form legal opinions. Rather, your determination merely needs to be whether or not you should involve law enforcement in the incident. As a general rule, if you think there may be a nexus for law enforcement to investigate and prosecute, then contact them. The law enforcement agency can then make their own determinations regarding the case.

18 US Code 1029 Fraud and Related Activity in Connection with Access Devices

This law is closely related to the 18 US Code 1030 already discussed. This law focuses on access devices, as the name suggests. It includes anyone who uses or traffics in counterfeit access devices. Counterfeit access device is defined as "any access device that is counterfeit, fictitious, altered, or forged, or an identifiable component of an access device or counterfeit access device". This is a rather broadly worded law.

18 US Code 1029 can be used to prosecute crimes that involve fake wireless access points, hacking into an access point, and related activities. It is not uncommon to see a case that uses both 1029 and 1030 to prosecute a given crime. This is another area that will require some initial triage on the part of the security professional conducting the incident response. The triage is to determine if law enforcement should be contacted.

An example of a case is the US vs Xavier Taylor, appealed to the Court of Appeals for the Eleventh Circuit court and decided in March 2016. In this case the defendant Xavier Taylor was alleged to have used stolen identities to add himself as an authorized user to the victim's credit card accounts and to open new accounts. This individual was charged with violation of 18 USC 1029 and 18 USC 1028A (which will be discussed later in this chapter). The defendant was convicted and sentenced to a total of 61 months of incarceration. This case illustrates how federal statutes are used in conjunction to prosecute cybercrimes.

Unlawful Access to Stored Communications: 18 U.S.C. § 2701

This act covers access to a facility through which electronic communication is provided or exceeding the access that was authorized. It is broadly written to apply to a range of offenses. Punishment can be up to 5 years in prison and fines for the first offense.

The actual wording of the statute is as follows:

Offense. —Except as provided in subsection (c) of this section whoever—

(1) intentionally accesses without authorization a facility through which an electronic communication service is provided; or
(2) intentionally exceeds an authorization to access that facility; and thereby obtains, alters, or prevents authorized access to a wire or electronic communication while it is in electronic storage in such system shall be punished as provided in subsection (b) of this section.

This law is used less frequently than the Computer Fraud and Abuse Act. However, it is written broadly enough to cover a number of acts. Primarily, the focus is on any facility, server, or device used to store electronic communications. It is sometimes the case that when employees leave a company, they seek to take information that they can use in competition with the company. This can include emails or other stored communications. Doing so is in direct violation of this law, and may also violate the Electronic Communications Privacy Act, 18 USC 2510, which will be discussed later in this chapter.

Identity Theft Enforcement and Restitution Act

This act was actually an extension of the Computer Fraud and Abuse Act. It was written in response to the growing threat of identity theft and the perceived inadequacy of existing laws. One of its most important provisions was to allow prosecution of computer fraud offenses for conduct not involving an interstate or foreign communication. This meant that purely domestic incidents occurring completely within one state were now prosecutable under federal law. Beyond that important provision, this act expanded the definition of cyber extortion to include threats to damage computer systems or steal data.

Another important aspect of this legislation was that it expanded identity theft laws to organizations. Prior to this, only natural persons could legally be considered victims of identity theft. Under this act organizations can also legally be considered victims of identity theft and fraud. This law also made it a criminal offense to conspire to commit computer fraud. Since this law was originally enacted, there have been additional laws to combat the problem of identity theft.

Gramm-Leach-Bliley Act

This law was passed in 1999 and was also known as the Financial Modernization Act. There are several aspects of the law that are not related to computer crimes and we will not be discussing here. The law did require financial institutions to develop a written information security plan. The plan must cover protecting the personal information of customers. The law was somewhat vague but was an attempt to legislate protections of personal data. The portion of the law regarding information security is called the Safeguards Rule. The Federal Trade Commission describes the Safeguards Rule as follows "The Safeguards Rule requires financial institutions under FTC jurisdiction to have measures in place to keep customer information secure. In addition to developing their own safeguards, companies covered by the Rule are responsible for taking steps to ensure that their affiliates and service providers safeguard customer information in their care." (https://www.ftc.gov/enforcement/rules/rulemaking-regulatory-reform-proceedings/safeguards-rule)

Identity Theft Laws

Identity theft has been a substantial problem for many years, and it shows no indications of abating. There are both state and federal laws that seek to address identity theft and related crimes. In this section we will look at the most commonly applied federal statutes.

18 U.S.C. § 1028A—Aggravated Identity Theft

Much like laws regarding aggravated assault, aggravated identity theft means a crime (in this case identity theft) is exacerbated by additional factors calling for more severe punishment. This law states that if someone is committing identity theft in relation to another felony such as wire fraud, illegally acquiring a firearm, embezzlement, immigration crimes, and several other crimes, then the penalties are increased.

The actual wording of the law is given here:

Whoever, during and in relation to any felony violation enumerated in subsection (c), knowingly transfers, possesses, or uses, without lawful authority, a means of identification of another person shall, in addition to the punishment provided for such felony, be sentenced to a term of imprisonment of 2 years.

(2) Terrorism offense.

Whoever, during and in relation to any felony violation enumerated in section 2332b(g)(5)(B), knowingly transfers, possesses, or uses, without lawful authority, a means of identification of another person or a false identification document shall, in addition to the punishment provided for such felony, be sentenced to a term of imprisonment of 5 years.

Essentially this law allows prosecutors to seek additional sentencing for identity theft when it is done relative to another crime. The US government published some interesting statistics about Aggravated Identity Theft (https://www.ussc.gov/sites/default/files/pdf/research-and-publications/quick-facts/Aggravated_Identity_Theft_FY17.pdf). Their date shows that there were 66,873 cases of identity theft reported to the United States Sentencing Commission in 2017. Of those, 1,043 involved convictions under 18 USC 1028A. The average age of the perpetrators when sentenced was 36, and the majority were US citizens. The majority of those convicted were also convicted of other cybercrimes.

The case of US vs Faraminan Eddys is an example case. Mr. Eddy's was accused and pled guilty to one count of conspiracy to traffic in unauthorized access devices (18 USC 1029) and three counts of aggravated identity theft (18 USC 1028A. He was sentence to 48 months for the violation of 18 USC 1029 and 24 months for the violations of 18 USC 1028A, for a total of 72 months. He appealed his sentence, that appeal was decided in 2017. What makes this case interesting from the cybersecurity perspective is that multiple offenses where involved. Identity theft is often concurrent with other crimes.

18 U.S.C. § 1028—Fraud and Related Activity in Connection with Identification Documents, Authentication Features, and Information

This law is not strictly speaking a computer law. It references identification documents. The goal is to target the manufacture, sale, or transfer of false identification documents. However, this is often done in conjunction with identity theft, thus it can be found in computer crime prosecutions. The law also includes 'authentication features' which is defined broadly as any "holgram, watermark, certification, symbol, code, image, sequence of numbers or letters…" thus providing a further nexus with computer crimes.

Copyright Laws

Copyrights are one method companies and individuals utilize to protect their intellectual property. Books, movies, software, and a wide range of other types of intellectual property may be subject to copyright protection. In addition to civil litigation, in some cases there are criminal penalties for violating copyrights or for facilitating the violation of copyrights.

No Electronic Theft Act of 1997

What is commonly called the No Electronic Theft Act of 1997, known also as the NET Act, was House Resolution 2265 signed into law by President Clinton on December 16, 1997. The purpose of this law is to provide law enforcement and prosecutors with the tools to fight copyright violations on the internet. Under this law, electronic copyright infringement can carry a maximum penalty of 3 years in prison and a $250,000 fine. This law made it a federal crime to reproduce, distribute, or share copies of electronic copyrighted works. This means not only software, but also music, videos, or electronic versions of printed material. Under this law it is a crime to distribute such copyrighted material, even if the distributer does so without any financial gain.

The law does require that the distribution be willful, and that the retail value of the copyrighted material exceed $1,000 in retail value. It is important for law enforcement and prosecutors to keep in mind that this law comes with a 5-years statute of limitations. In other words, the crime must be charged and prosecuted within 5 years of its commission.

While the distribution of copyrighted material is the key focus of this act, and the portion most often discussed in legal circles, it is not the only thing this law did. It also made it a criminal act to:

- Remove a copyright notice from an electronic product.
- Knowingly place a false copyright notice (in other words, to claim a copyright on something someone else already had copyright to).

Furthermore, the NET Act specifically addressed violation of copyrights on live musical or video performances. This means that it is a federal crime to record live performances without permission and then distribute such recordings.

In July 2016 a complaint was filed in the Northern District of Illinois accusing a Ukrainian man named Artem Vaulin of being the person behind the site Kickass Torrents. Among the crimes Mr. Vaulin was charged with were violations of copyright statutes in reference to copyrighted movies being trafficked through this site. Mr. Vaulin was apprehended by Polish authorities, but as of this writing is still fighting extradition.

Digital Millennium Copyright Act

This act, signed into law on October 28, 1998, frequently called the DMCA, focused primarily on methods for circumventing access control. The essential feature of this law is that it made it illegal to attempt to circumvent copy protection technologies. Manufacturers of CDs, DVDs, and other media frequently introduce technological measures that prevent unauthorized copying of the media in order to protect their copyrighted material.

This law did provide protection from prosecution for online providers, including internet service providers, if they adhered to certain measures. In other words, an ISP is not liable if one of its customers is using the ISP's service to violate the DMCA. This is common for many statutes. The only caveat is that the ISP is not liable, provided the ISP was unaware of the illegal activity and did not participate in the activity.

In addition to the protections for ISPs and other online providers, the law allows for the Library of Congress to issue specific and explicit exceptions to DCMA. Usually, these exemptions are granted when it is shown that a particular access-control technology has had a significant adverse effect on the ability of individuals to make non-infringing uses of copyrighted works. The specific exemption rules are revised every 3 years. A proposal for an exemption can be submitted by anyone to the Registrar of Copyrights.

This law has been controversial due to its broad nature, and its occasional use against people who might not otherwise be considered criminals. Princeton professor Mark Felton was threatened with legal action for writing a scientific paper explaining methods to circumvent certain watermarks used to establish copyright ownership. Felton initially withdrew the paper but then later presented it at the USENIX conference in 2001.

Laws Protecting Children

IT is an unfortunate fact that child predators make ready use of the internet. Trafficking in child pornography is the most obvious such example. There are a number of laws designed to combat this type of crime. Some are about shielding children from offensive and objectionable content, others about pursuing those who prey on children.

Children's Internet Protection Act

This bill was signed into law in 2000. The primary purpose of this bill was to require libraries and schools to filter content that children have access to. The law does require that libraries turn off the filter for adult patrons should they request it. The intent is simply to ensure that children are not exposed to pornographic or indecent material on computer systems supplied by the taxpayer. An internet safety policy must include technology protection measures to block or filter internet access to pictures that are: (a) obscene, (b) child pornography, or (c) harmful to minors (for computers that are accessed by minors).

Schools and libraries must also certify that, as part of their internet safety policy, they are educating minors about appropriate online behavior, including cyberbullying awareness and response and interacting with other individuals on social networking sites and in chat rooms. Schools subject to CIPA are required to adopt and enforce a policy to monitor online activities of minors. This means that in addition to filtering, the internet access must be monitored.

Schools and libraries subject to CIPA are required to adopt and implement a policy addressing: (a) access by minors to inappropriate matter on the internet; (b) the safety and security of minors when using electronic mail, chat rooms, and other forms of direct electronic communications; (c) unauthorized access, including so-called "hacking," and other unlawful activities by minors online; (d) unauthorized disclosure, use, and dissemination of personal information regarding minors; and (e) restricting minors' access to materials harmful to them.

18 U.S.C. § 1462—Importation or Transportation of Obscene Matters

This law is rather broadly written. The first issue with the law is one share by all attempts to curtail pornography, how to define "lewd", "obscene" and similar words. This law, however, has some additional proscriptions such as medicines for producing abortion. The specific wording in the statute is

(a) any obscene, lewd, lascivious, or filthy book, pamphlet, picture, motion-picture film, paper, letter, writing, print, or other matter of indecent character; or
(b) any obscene, lewd, lascivious, or filthy phonograph recording, electrical transcription, or other article or thing capable of producing sound; or
(c) any drug, medicine, article, or thing designed, adapted, or intended for producing abortion, or for any indecent or immoral use; or any written or printed card, letter, circular, book, pamphlet, advertisement, or notice of any kind giving information, directly or indirectly, where, how, or of whom, or by what means any of such mentioned articles, matters, or things may be obtained or made; or

The law is most often applied in cases involve child pornography as an additional charge to be added to the underlying child pornography charges. It is closely related to 18 USC 1465

18 U.S.C. § 1466A—Obscene Visual Representation of the Sexual Abuse of Children

This law has a specific purpose, to combat child pornography in all of its forms. The law specifically states that any visual depiction of any kind, regardless of medium that depicts a minor engaging in sexually explicit conduct is a crime
    The actual wording of the law is provided here:

(a) In general.—Any person who, in a circumstance described in subsection (d), knowingly produces, distributes, receives, or possesses with intent to distribute, a visual depiction of any kind, including a drawing, cartoon, sculpture, or painting, that—

   (1) depicts a minor engaging in sexually explicit conduct; and (B) is obscene; or
   (2) depicts an image that is, or appears to be, of a minor engaging in graphic bestiality, sadistic or masochistic abuse, or sexual intercourse, including genital-genital, oral-genital, anal-genital, or oral-anal, whether between persons of the same or opposite sex; and (B) lacks serious literary, artistic, political, or scientific value; or attempts or conspires to do so, shall be subject to the penalties provided in section 2252A(b)(1), including the penalties provided for cases involving a prior conviction.

One interesting element of this statute is the following phrase "It is not a required element of any offense under this section that the minor depicted actually exist." Therefore, cartoon, computer generated images, or other artificial images that do not depict an actual child, but rather an imaginary or fictitious child are still criminal acts.

18 U.S.C. § 2251—Sexual Exploitation of Children

This law involves the actual involvement of a child. It explicitly defines any person that is party to any part of such an act, is guilty. Even if that involvement merely involves transport of imagery created by another party. This statute carries substantial penalties. A single offense, including conspiracy to violate this statute, carry's a prison term of 15–30 years. The specific wording of this statute is given here:

(a) Any person who employs, uses, persuades, induces, entices, or coerces any minor to engage in, or who has a minor assist any other person to engage in, or who transports any minor in or affecting interstate or foreign commerce, or in any Territory or Possession of the United States, with the intent that such minor engage in, any sexually explicit conduct for the purpose of producing any visual depiction of such conduct or for the purpose of transmitting a live visual depiction of such conduct, shall be punished as provided under subsection (e), if such person knows or has reason to know that such visual depiction will be transported or transmitted using any means or facility of interstate or foreign commerce or in or affecting interstate or foreign commerce or mailed, if that visual depiction was produced or transmitted using materials that have been mailed, shipped, or transported in or affecting interstate or foreign commerce by any means, including by computer, or if such visual depiction has actually been transported or transmitted using any means or facility of interstate or foreign commerce or in or affecting interstate or foreign commerce or mailed.

(b) Any parent, legal guardian, or person having custody or control of a minor who knowingly permits such minor to engage in, or to assist any other person to engage in, sexually explicit conduct for the purpose of producing any visual depiction of such conduct or for the purpose of transmitting a live visual depiction of such conduct shall be punished as provided under subsection (e) of this section, if such parent, legal guardian, or person knows or has reason to know that such visual depiction will be transported or transmitted using any means or facility of interstate or foreign commerce or in or affecting interstate or foreign commerce or mailed, if that visual depiction was produced or transmitted using materials that have been mailed, shipped, or transported in or affecting interstate or foreign commerce by any means, including by computer, or if such visual depiction has actually been transported or transmitted using any means or facility of interstate or foreign commerce or in or affecting interstate or foreign commerce or mailed.

(c) (1) Any person who, in a circumstance described in paragraph (2), employs, uses, persuades, induces, entices, or coerces any minor to engage in, or who has a minor assist any other person to engage in, any sexually explicit conduct outside of the United States, its territories or possessions, for the purpose of producing any visual depiction of such conduct, shall be punished as provided under subsection (e).

(2) The circumstance referred to in paragraph (1) is that—

(A)   the person intends such visual depiction to be transported to the United States, its territories or possessions, by any means, including by using any means or facility of interstate or foreign commerce or mail; or

(B)   the person transports such visual depiction to the United States, its territories or possessions, by any means, including by using any means or facility of interstate or foreign commerce or mail.

(d)   (1) Any person who, in a circumstance described in paragraph (2), knowingly makes, prints, or publishes, or causes to be made, printed, or published, any notice or advertisement seeking or offering—

(A)   to receive, exchange, buy, produce, display, distribute, or reproduce, any visual depiction, if the production of such visual depiction involves the use of a minor engaging in sexually explicit conduct and such visual depiction is of such conduct; or

(B)   participation in any act of sexually explicit conduct by or with any minor for the purpose of producing a visual depiction of such conduct; shall be punished as provided under subsection (e).

18 U.S.C. § 2252B—Misleading Domain Names on the Internet [To Deceive Minors]

Federal statutes regarding misleading domain names may be seen like an overreach. However, as the name of this law indicates, there is a single purpose. To prevent purveyors of pornographic materials from using a domain name that might mislead a minor into visiting the website believing it has innocuous content. The actual working of the law is provided here:

(a)   Whoever knowingly uses a misleading domain name on the Internet with the intent to deceive a person into viewing material constituting obscenity shall be fined under this title or imprisoned not more than 2 years, or both.

(b)   Whoever knowingly uses a misleading domain name on the Internet with the intent to deceive a minor into viewing material that is harmful to minors on the Internet shall be fined under this title or imprisoned not more than 10 years, or both.

(c)   For the purposes of this section, a domain name that includes a word or words to indicate the sexual content of the site, such as "sex" or "porn", is not misleading.

(d)   For the purposes of this section, the term "material that is harmful to minors" means any communication, consisting of nudity, sex, or excretion, that, taken as a whole and with reference to its context—

(1)   predominantly appeals to a prurient interest of minors;

(2)   is patently offensive to prevailing standards in the adult community as a whole with respect to what is suitable material for minors; and

(3)   lacks serious literary, artistic, political, or scientific value for minors.

(e)   For the purposes of subsection (d), the term "sex" means acts of masturbation, sexual intercourse, or physical [1] contact with a person's genitals, or the condition of human male or female genitals when in a state of sexual stimulation or arousal.

18 U.S.C. § 2252C—Misleading Words or Digital Images on the Internet

This law is similar to the previous law, but is not restricted to minors. The essence of the law is that if adults wish to knowingly view legal pornographic materials, that is their choice. However, tricking people who do not wish to view such materials into visiting a website they may find offensive, is not acceptable. The actual working of the law is provided here:

(a)   In General. —
    Whoever knowingly embeds words or digital images into the source code of a website with the intent to deceive a person into viewing material constituting obscenity shall be fined under this title and imprisoned for not more than 10 years.
(b)   Minors. —
    Whoever knowingly embeds words or digital images into the source code of a website with the intent to deceive a minor into viewing material harmful to minors on the Internet shall be fined under this title and imprisoned for not more than 20 years.
(c)   Construction. —
    For the purposes of this section, a word or digital image that clearly indicates the sexual content of the site, such as "sex" or "porn", is not misleading.

State Laws

We have examined the common US Federal laws previously in this chapter. By the time of this writing, most states have their own individual laws regarding various cybercrimes. It is not feasible to examine each and every states laws in this chapter. A sample of such state laws has been selected to review in this section.

Texas Laws

Texas Penal Code Title 7 Chap. 33 directly addresses computer crimes (https://statutes.capitol.texas.gov/SOTWDocs/PE/htm/PE.33.htm). This is a broad-based law with a number of definitions tor everything from what access means to how to define property. This makes the law quite thorough and clear. Section 33.02 describes in detail the various breaches of computer security and what would constitute a felony vs a misdemeanor. some portions of the law are similar to certain federal standards, for example 33.021 addresses the online solicitation of a minor.

The law also covers interfering with electronic access (section 33.022). Perhaps the most unique element of the law is section 33.024 which describes 'unlawful decryption'. Decrypting someone's data without their permission can be a misdemeanor or felony depending on the dollar amount of the damages. The crime is defined as follows "A person commits an offense if the person intentionally decrypts encrypted private information through deception and without a legitimate business purpose."

The Texas statute also has a very thorough treatment of online impersonation (section 33.07). The law details why the person was creating the impersonation and specifically mentions use of social networking. Emails, texts, and other communication using another person's identity is also a crime.

California Laws

California has section 502 of Chap. 5 of the California Penal Code deals with larceny, and also addresses computer related crimes. It lists a number of items that are prohibited including:

> Knowingly and without permission uses the Internet domain name or profile of another individual, corporation, or entity in connection with the sending of one or more electronic mail messages or posts and thereby damages or causes damage to a computer, computer data, computer system, or computer network.

> Knowingly and without permission disrupts or causes the disruption of computer services or denies or causes the denial of computer services to an authorized user of a computer, computer system, or computer network.

> Knowingly and without permission uses or causes to be used computer services.

The specific coverage of impersonating an individual or company, along with explicitly calling out denial of service attacks is intriguing. The last one mentioned above is quite broad and would apply to many if not all computer crimes.

Rhode Island Laws

The state of Rhode Island has addressed a wide array of computer crimes via subsections of law 11-52. 11-52-3 addresses unauthorized access:

> Whoever, intentionally, without authorization, and for fraudulent or other illegal purposes, directly or indirectly, accesses, alters, damages, or destroys any computer, computer system, computer network, computer software, computer program, or data contained in a computer, computer system, computer program, or computer network shall be guilty of a felony.

The first thing of note in this law is that it makes these acts a felony. Under many state laws, unauthorized access is merely a misdemeanor. The other item of note is the broad language used. Phrases such as *"directly or indirectly"* and *"or other illegal purposes"* give prosecutors and law enforcement officials wide latitude.

11-52-4-1 deals with computer trespass and does so quite thoroughly.

(a) It shall be unlawful for any person to use a computer or computer network without authority and with the intent to:

   (1) Temporarily or permanently remove, halt, or otherwise disable any computer data, computer programs, or computer software from a computer or computer network;

   (2) Cause a computer to malfunction regardless of how long the malfunction persists;

   (3) Alter or erase any computer data, computer programs, or computer software;

   (4) Effect the creation or alteration of a financial instrument or of an electronic transfer of funds;

   (5) Cause physical injury to the property of another;

   (6) Make or cause to be made an unauthorized copy, in any form, including, but not limited to, any printed or electronic form of computer data, computer programs, or computer software residing in, communicated by, or produced by a computer or computer network;

   (7) Forge e-mail header information or other Internet routine information for the purpose of sending unsolicited bulk electronic mail through or into the facilities of an electronic mail service provider or its subscribers; or

   (8) To sell, give or otherwise distribute or possess with the intent to sell, give or distribute software which is designed to facilitate or enable the forgery of electronic mail header information or other Internet routing information for the purpose of sending unsolicited bulk electronic mail through or into the facilities of an electronic mail service provider or its subscribers.

(b) Nothing in this section shall be construed to interfere with or prohibit terms or conditions in a contract or license related to computers, computer data, computer networks, computer operations, computer programs, computer services, or computer software or to create any liability by reason of terms or conditions adopted by, or technical measures implemented by, a Rhode Island-based electronic mail service provider to prevent the transmission of unsolicited bulk electronic mail in violation of this chapter. Whoever violates this section shall be guilty of a felony and shall be subject to the penalties set forth in [ss] 11-52-2. If the value is five hundred dollars ($500) or less, then the person shall be guilty of a misdemeanor and may be punishable by imprisonment for a term not exceeding one year or by a fine of not more than one thousand dollars ($1,000) or both.

Under this law even temporarily disabling a computer program running on a system without authorization is a crime. Also listed as crimes in this law are altering or erasing any data or program, forging an email header, and distributing software that would facilitate the forging of an email header. This broad and thorough language is important because many times these techniques are used as part of a skillful hacker's process of compromising a target system. Rhode Island is one of the few states that address these issues directly and specifically.

Maine Laws

The Maine statute is of note primarily because it is rather limited. The law simply states

A person is guilty of criminal invasion of computer privacy if the person intentionally accesses any computer resource knowing that the person is not authorized to do so.

Now prior to this, the legislation does define key terms:

1. "Access" means to gain logical entry into, instruct, communicate with, store data in or retrieve data from any computer resource. [1989, c. 620, (NEW).]
2. "Computer" means an electronic, magnetic, optical, electrochemical, or other high-speed data processing device performing logical, arithmetic, or storage functions, and includes any data storage facility or communications facility directly related to or operating in conjunction with such device.

Now the definition of access is actually quite good. Logical entry has a well-defined meaning in the computer profession: It means the person need not be physically present. For example, when you log on to your bank Web site to check your balance, you have gained logical entry into the system. The problem here is the definition of computer and the fact that there are various types of unauthorized access. For example, purposefully using a password cracker to break into someone's bank account is one type of unauthorized access, accidently getting into someone's Yahoo email account because they forgot to log off when on a public terminal is quite another. This is the weakness of this particular legislation. It does cover all types of unauthorized access but provides no differentiation.

Alabama Consumer Identity Protection Act

The Alabama Consumer Identity Protection Act[10] is a very comprehensive piece of legislation. In fact, it is so extensive that it was split into several portions covering Alabama statues 13A-8-190 to 13A-8-201. The first thing this act did was to elevate the crime of identity theft from a misdemeanor to a class C felony. The classification of any crime is often a gauge of how seriously the legislature believes that crime is impacting society. In the case of identity theft, it would be difficult to argue that it does not have a very serious and growing impact on society. The Alabama legislature realized this and increased the legal severity with which this crime is treated.

The second interesting thing this law does is that it clearly defines what constitutes a violation of the statute. Let us look at the actual wording of this law:

(a) A person commits the crime of identity theft if, without the authorization, consent, or permission of the victim, and with the intent to defraud for his or her own benefit or the benefit of a third person, he or she does any of the following:

   (1) Obtains, records, or accesses identifying information that would assist in accessing financial resources, obtaining identification documents, or obtaining benefits of the victim.
   (2) Obtains goods or services through the use of identifying information of the victim.
   (3) Obtains identification documents in the victim's name.

(b) Identity theft is a Class C felony.
(c) This section shall not apply when a person obtains the identity of another person to misrepresent his or her age for the sole purpose of obtaining alcoholic beverages, tobacco, or another privilege denied to minors.
(d) Any prosecution brought pursuant to this article shall be commenced within 7 years after the commission of the offense.

This is very clear and unambiguous, and identifies exactly what actions violate this statute. This is a good example of a law against identity theft.

Florida Criminal Use of Personal Identification Information

The Florida statute 817.568 is an exemplary piece of legislation. The act does a very good job of clearly defining its scope, the terms used, and the consequences. More importantly, the law makes abundantly clear that the purpose of using someone's personal information is irrelevant. To quote the statute[11]:

Any person who willfully and without authorization fraudulently uses personal identification information concerning an individual without first obtaining that individual's consent commits a felony of the second degree.

What this means is that even if one does not use the personal identification information for financial purposes, it is still a felony. This is a very important point that not all identity theft laws address. As we discussed in Chaps. 1 and 2, while identity theft is often associated with economic crimes, it does not have to be. A perpetrator can use someone else's identity to discredit that person, to embarrass them, or to simply hide the perpetrator's own actions.

Of course, the law goes on to address identity theft for financial gain. In fact, it divides punishments based on the scope of the crime:

For crimes with damages between $5,000 and $50,000 in damages and/or affecting between 10 and 20 people, the penalty is a minimum of 3 years imprisonment.

For crimes with damages between $50,000 and $100,000 in damages and/or affecting between 20 and 30 people, the penalty is a minimum of 5 years imprisonment.

For crimes with damages in excess of $100,000 and/or effecting more than 30 people, the penalty is a minimum of 10 years imprisonment.

The wording in the law then explicitly states that these are minimum sentences and nothing in this law should be construed as prohibiting a court from imposing a greater sentence. These facts alone make the Florida statute worthy of examination. However, they have added an interesting item. Under this law, using personal identification information in the furtherance of harassment carries, in and of itself, additional penalties. The law states

Any person who willfully and without authorization possesses, uses, or attempts to use personal identification information concerning an individual without first obtaining that individual's consent, and who does so for the purpose of harassing that individual, commits the offense of harassment by use of personal identification information, which is a misdemeanor of the first degree.

The statute then continues on and specifies additional penalties and damages based on specific modifiers to the underlying crime. Specifically, the law states that if the identity information was gleaned from a public record, the crime just became elevated, as follows:

- A misdemeanor of the first degree is reclassified as a felony of the third degree.
- A felony of the third degree is reclassified as a felony of the second degree.
- A felony of the second degree is reclassified as a felony of the first degree.

I find this segment of this legislation particularly interesting and would hope all state legislatures would contemplate adding something similar to their own laws. Clearly a great deal of information is available online via public records, and to some extent this is quite helpful and useful. For example, if one is about to go into business with a partner, it is good to be able to easily find out if that partner has been involved in business litigation in the past, has filed bankruptcy, and so on. However, the Florida legislature, with this clause in the statute, is taking a rather harsh stance on those who misuse that information either in the course of identity theft or in the course of harassing someone.

New York Identity Theft Laws

The state of New York addresses identity theft via penal code 190.77–190.84[13], which covers identity theft and related charges. Statutes 190.77 through 190.80 define a variety of levels of identity theft crimes, including identity theft in the second and third degrees as well as aggravated identity theft. Identity theft in the third degree, the least serious offense under New York law, is defined as follows:

A person is guilty of identity theft in the third degree when he or she knowingly and with intent to defraud assumes the identity of another person by presenting himself or herself as that other person, or by acting as that other person or by using personal identifying information of that other person, and thereby:

1. obtains goods, money, property or services or uses credit in the name of such other person or causes financial loss to such person or to another person or persons.

Identity theft in the third degree is considered a class A misdemeanor. If the offense involves loss in excess of $2,000, New York statute 190.80 classifies that as identity theft in the first degree, which is a class D felony. 190.80-A further classifies identity theft wherein the victim is a member of the armed services who is deployed outside the continental United States with damages in excess of $500 as aggravated identity theft, which is also a class D felony. This fine tuning of identity theft laws is very interesting as it demonstrates that the New York legislature put some time and thought into these laws. It also demonstrates recognition on the part of New York that, like other crimes, identity theft can occur in varying degrees of severity.

The New York penal code 190.81 through 190.9–83 goes on to list three levels (first through third degree) of crimes for possession of personal identifying information without authorization. The law states:

A person is guilty of unlawful possession of personal identification information in the third degree when he or she knowingly possesses a person's financial services account number or code, savings account number or code, checking account number or code, brokerage account number or code, credit card account number or code, debit card number or code, automated teller machine number or code, personal identification number, mother's maiden name, computer system password, electronic signature or unique biometric data that is a fingerprint, voice print, retinal image or iris image of another person knowing such information is intended to be used in furtherance of the commission of a crime defined in this chapter.

This is important because it makes it a crime to simply possess information that one could use to commit fraud. This has a similar effect to Idaho's legislation in that it makes various aspects of identity theft to be individual crimes. It also gives law enforcement the opportunity to prosecute a case in which the fraud aspect may not be provable, but the possession of personal identification information is. For example, if in the course of an investigation a law enforcement officer discovers a suspect's computer has personal identifying information for 20 people, that suspect can be charged with 20 counts under this law. This holds true even if any use of this information in the commission of a fraud cannot be proven.

Conclusions

This chapter should have provided you a general overview of United States computer crime laws. The goal has been to provide you a basic understanding of the relevant computer crime laws. As a cybersecurity professional you should have some understanding of computer criminal law.

Chapter 5
Network Management

Lo'ai Tawalbeh

Introduction

The importance of networking and distributed processing systems in organizations is growing each day and larger and more complex networks are emerging. The distributed applications and associated resources become important as the network grows. The scale of the network increases and at the same time the probability of occurrence of network issues also increase which might result in destabilize the network and degrade the network performance.

Human effort is not enough to manage large networks due to the large level of complexity, and so there is an increasing need for automated network management tools. The demand for such tools has increased over the last few years since the decentralization of network services increased the complexity of network management tasks where in many networked information systems, the resources that need to be managed are far away from network management personnel [1].

This chapter discusses network management systems requirements, functions, existing network management strategies, and future trends. For effective network management, the requirements for network management and the current behavior of a network should be understood. A network management system can successfully manage a combined LAN/WAN environment. The system consists of data control and gathering tools which are integrated with the network software and hardware. The network management system's architecture will be examined together with the standardized network management software package.

© The Editor(s) (if applicable) and The Author(s), under exclusive license to Springer Nature Switzerland AG 2020
I. Alsmadi et al., *The NICE Cyber Security Framework*,
https://doi.org/10.1007/978-3-030-41987-5_5

Network Management Requirements

A network management system refers to a set of network control and monitoring tools integrated to perform network management activities. The system usually has a single operator interface and a set of commands to accomplish the tasks. Software and hardware equipment is integrated into the existing equipment. A network management system comprises of incremental software and hardware additions coordinating with the existing network components. There are many categories of network management systems based on the activities they perform, and so the for each management task the requirements will be different accordingly [2].

Fault Management Overview

In order to ensure proper operation of today's complex networks, the behavior of critical components should be monitored. Whenever a fault occurs, it is important to initiate a procedure for determining the type and the extent of the fault. Secondly, the section of the network having the fault should be isolated from the rest of the network. This will allow the healthy part of the network to operate without interferences. The network is the modified or reconfigured to ensure proper operation in the absence of the failed component. Finally, the failed components should be replaced or repaired to bring the network back to the normal operation [3].

The definition of fault is important under fault management. Faults should be differentiated from errors. Faults are abnormal conditions which usually demand a management action in order to repair and are characterized by excessive errors and operational failures. If a communication line, for instance, is cut, signals cannot be transmitted to anywhere within the network and so high bit error rate may result. On the other side, errors, such as single-bit errors, occur occasionally and cannot be considered as faults. Error control techniques can be used to compensate for errors.

User Requirements

Users anticipate reliable and fast problem-solving. In most cases, occasional errors can be tolerated by users. However, when such errors occur, the users expect a notification followed by immediate correction of the error. In order to achieve high level of fault resolution, there is a need for reliable and rapid fault detection and troubleshooting mechanism [4]. A high degree of tolerance can be achieved by using alternative routes and installing redundant components. To raise the level of network reliability, redundancy in fault management can be introduced. Users need to be continuously updated on the status of the network. They should be informed of any disruptive maintenance. Users should be assured of correct network operation. This is achieved by the use of alerts, logs or statistics. A fault management service should ensure that there no new issues after fault correction and system restoration. It should give an assurance that a fault has been completely resolved. This is referred to as problem control and tracking. Moreover, network performance should not be affected by fault management [5].

Accounting Management Overview

In most enterprise networks, the individual project accounts and divisions are charged based on the usage of network services. This does not involve actual cash transfer. There are internal accounting procedures used by the network manager in order to track the usage of network resources by each user. This assists the managers in monitoring the abuse of access privileges. Network managers can also identify activities carried out by network users which make a network inefficient and recommend procedures that can improve the overall network performance. Accounting management helps network managers to obtain adequate details necessary in planning network growth [6].

User Requirements

The network manager has the task of specifying the types of information which will be recorded across the nodes. He should also set the interval between successive transmission of the recorded information to the management nodes. The algorithms used in the calculation of charges should also be specified. A network manager should be in a position to generate accounting reports [7]. Access to accounting information can be regulated by introducing a user verification feature in the accounting facility. This will handle unauthorized access and manipulation of information.

Configuration and Name Management Overview

The existing data communication networks are made up of logical subsystems and components whose configuration can be modified to perform different tasks. A device, for instance, can be configured as an end device or as a router. The configuration manager chooses suitable values, attributes and software for a device once its function has been determined. Configuration management accomplishes network initialization and final network shut-down. It is also initiated in maintenance, addition and update of relationships between components and their status across the network [8].

User Requirements

Configuration management is concerned with starting up and shutting down operations across the network. Automatic initiation of these operations is desirable in some network components. Network managers should be able to recognize network components and determine their appropriate connectivity. When network configuration is carried out on a regular basis without a change in the set of resource attributes, mechanisms for defining and modifying default attributes should be introduced [9]. Also, the means of loading the predefined sets of attributes to the network components should be devised.

The network manager should be in a position to alter network components' connectivity whenever a user requests for a change. Networks are reconfigured after performance evaluation or during network fault recovery, upgrade or when introducing more security checks. Users expect regular update on the status of network components and resources. They should be informed about all network modifications. Configuration report may be generated on regular basis or when requested. Users usually inquire about the status of attributes and resources after reconfiguration. Network managers ensure that only authorized operators can control and manage operations across a network.

Performance Management Overview

The current data communications networks are made up of different components that coordinate and share information. There are circumstances where communication performance is kept within certain limits in order to maintain effectiveness of an application. In a computer network, performance management has two main functional categories: controlling and monitoring functions. The activities are tracked on the network by the monitoring function. The controlling function is used to make adjustments aimed at improvement of network performance. Some of the issues which affect network performance include excessive traffic, bottlenecks and poor response time. To handle such issues, network manager should set resources that can be monitored. This will also allow him track network performance level. The appropriate metrics and values are compared with the existing network resources in order to understand the performance of a network. In order to accurately determine the level of network performance, many network resources should be monitored. The information collected is analyzed and used to set prescribed values. The network manager understands the behavior of the network as he collects and analyzes more information [10].

User Requirements

A user can determine a particular application of a network after evaluating the reliability and the worst response time of the network services. One should adequate details on performance in order to make an appropriate response to specific user queries. Network users expect good response time from their applications. Performance statistics are useful to network managers. The statics helps the in network maintenance, management and planning. Using performance statistics, potential bottlenecks can be identified before problems occur. It enables networks managers to initiate correction actions before the end users experience the impact of a particular network issue. Some of the actions taken may include traffic load redistribution or balancing by changing the routing table. A load can be accumulated in one area when a bottleneck is identified. Moreover, the capacity planning, in the long run, is based on the performance information [11].

Security Management Overview

The generation, distribution, and storage of encryption keys is handled under security management. The maintenance and distribution of passwords and other access control information are important. Moreover, security management also deals with the control and monitoring access to networks. The information obtained from network management is used to ensure the security of the users' data [12]. Adding to that Logs are important in network security management. Therefore, the examination, collection and storage of logs play an important role in security. Control of logging facilities is involved in security management.

User Requirements

Security management offers facilities for protecting user information and network resources. Only authorized should be given access to network security facilities. The users want to be assured of adequate security policies and proper management of security facilities that will make their data more protected and secure [13].

Network Management Systems Examples

The software is used to perform network management activities within the communications processors and the host computers. Communications processors include routers, front-end processors and bridges. The design of network management systems allows network administrators to view the whole network as a unified architecture, with labels and addresses given to each point and the attributes of all the ele-

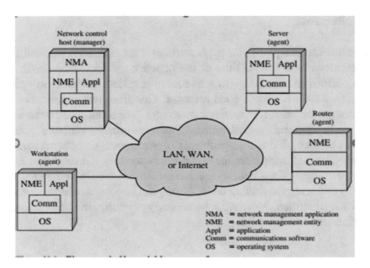

Fig. 5.1 Elements of a NMS

ments. The network control center receives regular update from the active elements in the network.

Figure 5.1 shows a network management system's architecture. Every network node has a set of software dedicated to network management activities. This is referred to as a network management entity (NME) as can be seen from Fig. 5.1. The roles of NME include a collection of statistics on communications and network-related events, local storage of statistics, generating artificial traffic during performance tests and responding to network control center commands. The network control center can command network nodes to transmit collected statistics, modify parameters or give the status of information [6].

In any network, at least one host should be designated as the manager or network control host. Apart from the NME software, there is a set of software referred to as network management application (NMA) in the network control host. Network management application has an interface used by network administrators to manage networks. The application accepts user commands and issues it to NMEs across the network. This is facilitated by application-level network management protocol whose architecture resembles that of a distributed application.

Network nodes are part of a network management system and include a NME which are sometimes referred to as an agent. User applications are supported by agents situated at the end systems. They are also present at the nodes which offer communications services such as routers, bridges and cluster controllers [7].

The network control host controls and shares information with NME as shown in Fig. 5.1. More than one network hosts are needed to achieve high availability of the network management function. Under normal circumstances, network control is accomplished by one of the centers and the other centers collect data or remain idle. The backup systems are used when the primary network control host fails.

There another management protocol which is the simple network management protocol (SNMP) which was initially developed as a network management tool for networks operating on TCP/IP. However, it has been its operations have been extended across all kinds of networking environments. The term simple network management protocol (SNMP) refers to a set of network management specifications which include the protocol itself, the database definition and related concepts. The key elements of SNMP include a network management protocol, management station, management information base and agent [14].

Challenges and Problems

Network managers are facing many network management challenges in daily basis. One of the main challenges is the need for maximizing networks' productivity without incurring high costs of network ownership. Network ownership costs include staff training, implementation, operation and administration costs. There are many network management solutions and tools available in the market, and these tools change with the business needs of the network which will help organizations to handle critical and complex systems [10].

Some examples of these tools include HP Open View, Ciscoworks, CA Unicenter, IBM Tivoli, and Novell ZENwork [15]. All of these tools use the centralized architecture and SNMP protocol to manage the network. Network management tools have advanced to span heterogeneous networks, protocols and equipment in a complex communications environment of voice, video and data. Most of the products accomplish this by being out of the box, with little or no customization.

The most commonly used NMS framework in a heterogeneous IT environment is HP OpenView [16]. This framework demands a lot of administrative overhead. Its message browser and the NNM map are old-fashioned. Working with OVO agents is not easy and systems need to be restarted when the agents fail to function. The process of installing OVO agents is tedious and time-consuming. The costs of this solution are high and may not be practical for small organizations. It is also complex to implement this solution.

CiscoWorks (CW) is a recommendable tool when the network is primarily Cisco-based. With this tool, such network can be easily managed. However, CiscoWorks (CW) cannot recognize all the non-Cisco devices. There are times when its response is very poor. CiscoWorks (CW), is based on Cisco Discovery Protocol and one needs CDP knowledge in order to troubleshoot and implement this product [17].

CA Unicenter Systems and Network Management allow clients to maintain high performance and availability of critical services by offering an integrated view of notification and events across organizations' entire IT infrastructure. Network facilitates coordination of application components running in various servers and platforms [18]. The performance and the availability of applications relies on the responsiveness and functioning of a network. CA Unicenter is expensive and suitable for large businesses. It demands a deep knowledge on protocols and technology. This increases the complexity of the tool.

SolarWinds Orion Network Monitoring tool is based on SNMP. Its interface is user-friendly and can be easily installed. However, its capability is limited to Fault Management. Its full functionality can only be realized by combining it with other products. This is expensive and small businesses may not afford. This tool needs to be customized to understand and unlock all the features [7].

Nagois is one of the Linux-based Open Source tool widely used in Network Management. It is easy and simple to install this tool. However, it requires manual configuration. Nagois has a poor user interface. Only individuals with networking and Linux administrator skills can use it. Despite the fact that Nagois is an open source and freely available, its complexity does not make it a good NMS [15].

Current Network Management Strategies

The emerging network management solutions take into account the cost and complexity issues associated with the current NMS. Organizations desire a network management tool with a simple GUI and easy to install. This will reduce the costs of NMS and improve data center management experience. The licensing of a network management tool should be simple and should support Internet-enabled technology. Also, the entire network should be managed from a single interface [11].

For a proactive NMS, the deployment strategies begin with the fault management activities and then slowly progress to the performance management. This is followed by configuration management and finally the proactive network management. Network configuration can only be refined after understanding the entire enterprise network topology [7]. In order to monitor the performance, fault and the configuration of a network, the parameters to be examined should be clearly defined.

Based on that, choosing a network management strategy requires an in-depth review of the organizations' expectations and goals. There are certain factors that need to be taking in consideration when strategizing network management such as the network type, company size, and financial impact of outages and events. There are different functions of network management that include the support of users and hardware, support of services, and service level agreements. Every organization from well-established enterprise organizations with thousands of employees to start up entrepreneurial organizations with a hand full of employees will follow a similar methodology in choosing the network management strategy that best suits their organization in the current phase of business. Once the strategy is decided on, it will then be planned, then implemented. Strategy planning does not end there or ever. It is revisited at least once a year to ensure you are still utilizing the best structure given any changes in expectations, goals, and organization size.

Network Type

Organizations network type play a significant role in network management strategies. Along with the organizational expectations and goals, network type will also play a significant role in dictating the network management strategies [19]. We will discuss some of the most common network types in current use.

Local Area Networks (LAN) is one of the simplest types of networks in current use outside of a peer to peer network. A local area network consists of a group computer and other network devices such as a printer and scanner. They are all centrally connected to a Router that is also providing the function of a switch. The router will typically be connected to a device providing internet connectivity to the group of devices on the network. These simple networks do not require very high levels of management. In many cases, there is no dedicated staff overlooking the performance or maintenance of the system.

Metropolitan Area Network (MAN) is a network that spans a city-wide area. Metropolitan Area Networks are a combination of several Local Area Networks. Each with a group of computer and network devices connected to a router with an external connection to the internet. From that connection to the internet interconnectivity of the individual routers at each office or home office is created. Metropolitan Area Networks is roughly the network size when organizations start to hire dedicated network management personal.

Wide Area Network (WAN) is similar to a Metropolitan Area Network in regards to being a combination of interconnected through internet backhauls, but a Wide Area Network is inter-connects several larger individual networks such as Enterprise networks, campuses, government facilities or even cities.

Wireless Local Area Networks (WLAN) are still broken out as their own network type in most textbooks and exams, but in the actual use of them, they are more of an overlay to every other type of network. This has been the faster-growing type of network. The growth has gone beyond their use and popularity. Technology growth in speed, throughput, distance, and security is what has made this into a lot of organizations go to network. School, hospitals, enterprise corporations, and small mom and pop ones along with almost every household in the United States, has deployed a wireless network of some sorts in the past decade. Just like the other networks above the complexity level of the network plays a large factor in the network management strategy for it. A Small Office Home Office (SOHO) Network with wireless connectivity will be much less complex than a Hospital Network providing Wireless VoIP phone systems or Bluetooth triangulation to track medicine charts. As the complexity of the network grows, the more specialized skill set and equipment your network management team will require. Some of the tools utilized for this are RF Spectrum Analyzers and RF Site Survey tools.

Enterprise Private Network (EPN) is larger and more complicated of the networks we are discussing. An Enterprise Private Network serves thousands of users at any of the given properties interconnected. Large corporations like AT&T, USAA, and many other top companies are great examples of Enterprise Private Networks. Not

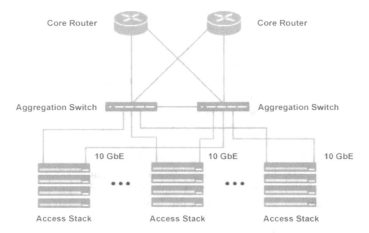

Fig. 5.2 Enterprise private network

only do these networks serve thousands of user, but they also manage a very large amount of data. As can be seen from Fig. 5.2, Enterprise networks begin to increase in complexity and redundancy.

Functions of Network Management

There are many functions performed by the network management systems. Regardless of the size or complexity of your network, there are many features and functions of network management that all organizations should consider as they implement network management. Network Dashboard plays a large role in the function of network management. A dashboard is a centralized location where the health of your network can be monitored. You, as the networking professional or a modest user who is overseeing the network, will gain various benefits from the implementations of a solid dashboard. Dashboards may be vendor-specific or vender neutral. Basic overview of the network devices, configurations, and performance are very instrumental when working on a network performance issue. You are able to view current status as well as logs of alerts that may have been recorded from prior incidents. This will help your team identify current issues or perhaps trends in the network. This information will allow your team to make a well-informed decision when troubleshooting. You will identify and resolve issues quicker than when working on individual hardware without the overall view.

Up/down monitoring and auto alerts go hand and hand. Your dashboard will utilize protocols such as SNMP or even a simple ping to identify when a device is no longer responsive [15]. The visual representation of that device will go from green to red. At this point, you would want some form of automation to take over. A good example would be to have your IP based devices report in every 5 min when a device is deemed as being offline for 5 min the color representation of that device changes to red. If the device missed two additional heartbeats for a total of 15 min of being offline you can send out an automated alert to a designated team at the property where the down device is based from or to your network management team. This will serve as an initial response requesting for someone to get eyes on the hardware and perform a reboot of the device. If the dashboard finds the device has recovered, then an all-clear notification goes out. If the device does not recover in another 15 min, then an auto ticket is created into the groups ticketing system for human interaction. There are many ways to set up these triggers; an organizations staffing and tolerance for downtime will dictate how tight or loose the automation are set to individual outages.

Incident management in a ticketing system is another important function of network management. Ticketing systems allow your team to track all open incidents and to prioritize them based on the strategy your organization uses to deem the which is most important. Some popular strategize for ticket prioritization are assigning a level of urgency to each ticket the sorting the queue with the most urgent on top. Another popular way is FIFO first in first out; with this method, each ticket importance is weighed equally. The system will sort them by age, and your staff will always pull the oldest ticket from the queue. Each method has there pros and cons; you and your team will need to find which method serves your customer best. Yes, the end-users of the network you manage are your customers, even if you all work for the same organization.

Network Security can also be part of your network management strategy [12]. I say it can be because, in today world, many organizations have broken out security out to be their organization within the company with separate leadership and budgets. A big factor in this decision is the sensitivity of the data you work with and the size of the organization. Security teams are then broken down even further into specialized teams. Some part of the team may be responsible for users permissions, while other monitor databases, or external threats. A smaller organization may outsource the work or perhaps have a jack of all trades multi-hat wearer leading a small or one-person team.

Identifying who is responsible for your network services is also key. Are these set them and forget them services or do you implement some sort of review and improvement rhythm? Basic services such as DHCP, DNS, NAT, Authentication, Bandwidth Management, and print services all need to have a responsible team identified in order not to send the company scrambling when one fails. Unlike up/down monitoring these are not as black and white to identify when there is an issue [16].

The end-user support team is the team who helps with the user level issues that arise on the day to day operations in an organization. Not all organization or networks require a dedicated team to support end-user. But those that do grow to that size see a large benefit by from having a dedicated to the team to assist end-user. Problems such as email configuration settings, VoIP phone issues, print server issues, application support, and flat out user ignorance are all items that can be addressed by the end-user support team. When growing a support team, you need to focus on the moral character of the people you bring on board and not just the technical qualities. Just like in a classroom setting, there are a few students that just naturally gifted and learn IT concepts with no struggle, but they may lack motivation or maturity. This lack then drives them to underperform and perhaps scrape by and pass the class. Those students will hit the workforce with the exact same work ethics and be poor performers. On the other hand, there are the students that stay in the median average of classes but fight night and day to the grade they get. They show a great work ethic and desire and ability to learn. Those are the employees you want to join your team. After performing the job for a while they will master it and be hungry to learn more and advance within the company.

As you implement you, end-user support team, not only because you have to find the best team members that fit the roles technical abilities but also fit the companies values. You will also have to ensure you staff your team properly to provide adequate coverage during peak time but also ensure you have coverage that mimics the other departments' schedule. For instance, if you are supporting an organization that runs a 24/7 operation, your team will need to have the plan to support the overnight team along with the day team. If the volume during the day is expected to be triple that of the night, your staffing will need to reflect those patterns as well. One way to ensure you are providing adequate staffing is to implement a service level agreement or a service level target then staff your team to ensure you meet those. A service level agreement is a minimum standard of service that is set form the support group to the group you are supporting [20]. For example, you may implement 80% of all the call are answered in under a minute as a service level. You then staff to ensure that you can meet that metric.

Network Support Team is a separate support team. The responsibilities of this team may overlap with some of the responsibilities of the end-user support team. The network support team responsibilities go deeper into the technical areas such as the services provided to keep the network running. The technical capabilities of this team should be higher than your user support team. The network support team will need prior experience in managing network resources and hardware. They may be required to perform hardware upgrades and swaps. Perhaps some physical wire runs or fiber terminations, swapping and programming access points.

Another important issue to address in network management is the staffing options. In other words, it is about deciding how the staff will play a significant factor in your overall organizational strategy (hiring direct employees or vetting contractors).

In house teams are direct hired dedicated full-time employees. This is one option to fullfill the task of network management. An in-house team will typically have an exceptional understanding of the company. This will help the team as they navigate the inter-workings of other departments. The organization will have complete over-sight on who is chosen to join the team allow it to shape the culture of it. Processes established will also likely fit the values of the company. There are some limitations when using a fully in-house team. Lack of scalability is usually the top detractor from this type of team. Providing the correct coverage hours is another.

Outsourcing is another solution for providing network management. You typically see this in a growing company as they work through the growing pains. Perhaps the

Fig. 5.3 Payroll outsourcing by company size

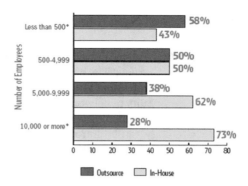

* Total percentages exceed 100% due to rounding.

Source: Goldman Sachs; American Payroll Association.

sales and operations team go through a quick growing season. An outsourced group would be brought on to ensure the proper support and SLA's are kept to keep the company moving. The outsourced teams are able to scale up and down well as the organization needs [20]. In doing so, they also free up the organizations key player to focus on the most important task at hand. The internal rock star team will not be bogged down with repetitive daily task so they will be able to work on the items that will help drive the company in the right direction. There is some risk that does present themselves when going with an outsourced team. These risks have to be closely guarded to ensure they do not become a problem. The security risk is the top risk, can you trust your data in the hands of a third party. Ensure the outsource partner adheres to the same security protocols as your dedicated staff. Culture mismatch is another threat when using a group of shared resources. As seen in Fig. 5.3, the larger an organization the less outsourcing they use.

The third choice for staffing the network management team is co-sources. Co-sourced is a hybrid model that incorporates the best of both worlds, in-house and outsourced. This allows the best scaling of all the models.

New/Future Trends

There are many new and future trends in network management driven by the rapid advances in communication technologies. The next subsections address some of these trends.

Technology of Data Acquisition

Polling and trap technology will be used to achieve network management by acquiring and processing data. The increase in the network complexity and capacity, results in an increase in the amount of data that can be acquired from the network. This demands high quality of network transmission and better network management. Such situation cannot be handled by the traditional data acquisition technology. Mobile agent technology can be introduced to break through such bottleneck. This technology exhibits features of initiative, independent, cooperation and interaction that ordinary agents have [21].

The intelligence and portability of mobile agent technology can reduce the capacity of data across a network. This technology removes the various step of data processing. Therefore, the load supported by the management station will be greatly reduced. Large heterogeneous distributed networks can be well managed using transparent, good portable and highly flexible mobile agent

Technology of Data Management and Display

In traditional network management technology, the remote network equipment is managed from the administration console. The concepts WBM technology, Xml technology and Portal technology can be used to implement a more intuitive and convenient network management techniques. This will allow network administrators with different permission levels to manage and view data from different parts of the world.

Network Management Based on Web

Web-Based Management technology combines that advantages of network management technology and Web function. This gives a network administrator a powerful that cannot be paralleled the traditional tools. WBM technology gives administrators the ability to configure and control network via a web browser. This technology will transform network management and will improve steps towards effective network management [15].

There are two ways that web-based network management can be achieved. One of the ways is using agent model where internals workstation is combined with a Web server which communicates with the terminal equipment. A communication channels is set up by browser users via HTTP protocol. The agent, on the other hand, uses SNMP protocol to set up a communication channel with terminal equipment via SNMP protocol (Fig. 5.4). The embedded system model can also be used to realize WBM technology. The web property, in this model, is embedded into the network

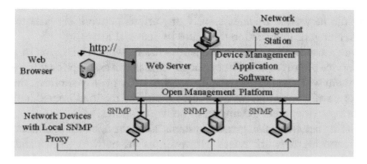

Fig. 5.4 Proxy model

devices which have their own IP addresses. A browser can be used to manage and access network devices [15].

The two models have their own respective advantages. They can be combined to achieve better network management together. The agent model introduces the advantages of mobility and flexibility while preserving the initial advantages of the network management system. The embedded system model introduces an aspect of graphical management technology. This delivers a simple user interface which facilitates easily network management.

XML Technology

Extensible Markup Language (XML) technology can be used to unify the current multiple management interfaces across the network system. The system offers a standard information source described in a form that can be easily read and understood by different systems and users. Therefore, the network application is universal and interoperable. An additional advantage of this technology is that it is very flexible. The technology increases the reliability of interactive operation between network devices and management system. The markup language used in the development of new management system is simple. The cost of implementation of such management system is low [22].

Summary

The advancements in network technology demand fast and effective management. The existing network management tools are not efficient. There is still a lot of research and development which need to be done in network management. Network connectivity has expanded rapidly over the last few decades. This expansion is accompanied by an increase in the number of networks and the volume of data traversing the networks. Businesses rely on various network services to run their operations. Network interruption translates to financial losses.

There are various classes of traffic across the network. Each type of traffic has different service requirements. The applications make an assumption that the network will offer them with the best service. In order to offer quality services, the network should operate efficiently. Network failures should be rectified immediately. Clients and network providers feel the impact of network failures.

The existing network management systems assist the network operators in detecting and diagnosing network issues. However, as the network grows and becomes more complex, network's self-management is desirable. A network should be able to detect and diagnose its problems. This is made possible by introducing intelligent network management systems.

References

1. Abeck, S. (2009). *Network management know it all*. Morgan Kaufmann.
2. Hashish, Sonia, & Tawalbeh, Hala. (2017). Quality of service requirements and challenges in generic WSN infrastructures. *Procedia Computer Science, 109,* 1116–1121.
3. Farrel, A. (2008). *Network quality of service know it all*. Elsevier.
4. Ding, J. (2016). *Advances in network management*. Auerbach Publications.
5. Saldamli, G., Mishra, H., Ravi, N., Kodati, R. R., Kuntamukkala, S. A., & Tawalbeh, L. (2019). Improving link failure recovery and congestion control in SDNs. In *2019 10th International Conference on Information and Communication Systems (ICICS)*, pp. 30–35. IEEE.
6. Clemm, A. (2006). *Network management fundamentals*. Cisco Press.
7. Claise, B., & Wolter, R. (2006). *Network management: Accounting and performance strategies*. Cisco Press.
8. Khan, R., Khan, S. U., Zaheer, R., & Babar, M. I. (2013). An efficient network monitoring and management system. *International Journal of Information and Electronics Engineering, 3*(1), 122.
9. Bhatt, S., Lo'ai, A. T., Chhetri, P., & Bhatt, P. (2019). Authorizations in cloud-based internet of things: Current trends and use cases. In *2019 Fourth International Conference on Fog and Mobile Edge Computing (FMEC)*, pp. 241–246. IEEE.
10. Kim, Hyojoon, & Feamster, Nick. (2013). Improving network management with software defined networking. *IEEE Communications Magazine, 51*(2), 114–119.
11. Curtis, A. R., Mogul, J. C., Tourrilhes, J., Yalagandula, P., Sharma, P., & Banerjee, S. (2011). DevoFlow: Scaling flow management for high-performance networks. In *ACM SIGCOMM Computer Communication Review*, 41, no. 4, pp. 254–265. ACM.
12. Jararweh, Y., Bany Salameh, H. A., Alturani, A., Tawalbeh, L., & Song, H. (2018). Anomaly-based framework for detecting dynamic spectrum access attacks in cognitive radio networks. *Telecommunication Systems, 67*(2), 217–229.
13. Tawalbeh, L., Al-Qassas, R. S., Darwazeh, N. S., Jararweh, Y., & AlDosari, F. (2015). Secure and efficient cloud computing framework. In *2015 International Conference on Cloud and Autonomic Computing*, pp. 291–295. IEEE.
14. Affandi, A., Riyanto, D., Pratomo, I., & Kusrahardjo, G. (2015). Design and implementation fast response system monitoring server using Simple Network Management Protocol (SNMP). In *2015 International Seminar on Intelligent Technology and Its Applications (ISITIA)*, pp. 385–390. IEEE.

15. Guimarães, V. T., dos Santos, G. L., da Cunha Rodrigues, G., Granville, L. Z., & Tarouco, L. M. R. (2014). A collaborative solution for SNMP traces visualization. In *The International Conference on Information Networking 2014 (ICOIN2014)*, pp. 458–463. IEEE.
16. Zitello, T., Williams, D., & Weber, P. (2003). *HP openview system administration handbook: Network node manager, customer views, service information portal, openview operations.* Prentice Hall PTR.
17. Rao, Umesh Hodeghatta. (2011). Challenges of implementing network management solution. *International Journal of Distributed and Parallel Systems, 2*(5), 67.
18. Hong, J. W. K., Kong, J. Y., Yun, T. H., Kim, J. S., Park, J. T., & Baek, J. W. (1997). Web-based intranet services and network management. *IEEE Communications Magazine, 35*(10), 100–110.
19. Jararweh, Yaser, Ababneh, Huda, Alhammouri, Mohammad, & Tawalbeh, Lo'ai. (2015). Energy efficient multi-level network resources management in cloud computing data centers. *Journal of Networks, 10*(5), 273.
20. Bahwaireth, K., & Tawalbeh, L. (2016). Cooperative models in cloud and mobile cloud computing. In *2016 23rd International Conference on Telecommunications (ICT)*, pp. 1–4. IEEE.
21. Lo'ai, A. T., & Saldamli. (2019). Reconsidering big data security and privacy in cloud and mobile cloud systems. *Journal of King Saud University-Computer and Information Sciences.*
22. Sharma, S., Goyal, S. B., Shandliya, R., Samadhiya, D. (2012). Towards XML interoperability. In *Advances in Computer Science, Engineering & Applications*, pp. 1035–1043. Springer, Berlin, Heidelberg.

Chapter 6
Risk Management

Lo'ai Tawalbeh

Introduction

Risk Management is an important security measure that organizations take to protect current assets. Since hackers are always looking for new and more practical methods to obtain access to you or your company's most sensitive data, the practice of risk management is perpetual and evolving. The process of developing a risk managementm strategy occurs in a four-step process: identifying risks, assessing the danger, prioritizing the risk and appropriately addressing the risk. It is important to note that each of these steps in the four-step process occurs independently of one another, but it is not impossible for any of these steps to coincide with one another. For example, it is possible for a company to assess the danger of a potential risk while also prioritizing the risk in a situation. This is seen in situations in which companies must protect themselves against threats such as viruses (where the risk is known), forcing organizations to prioritize the risk so that they can protect against potential risks associated with the organization.

The main goal of risk management is to always be one step ahead of the issue so that it never actually evolves into a problem. "*Organizations that align security with their strategic business objectives can drive business success with risk mitigation*" [1]. Aligning security with the business plan ensures that companies are goal oriented and only risks that potentially affect the prosperity of the business are addressed. While there are many risks to managing a business, especially IT related businesses, not all risks are pertinent to the needs of a particular organization. For example, store-front businesses are constantly faced with the potential of being burglarized. To prevent burglarization, businesses can analyze their risks and take preventative measures (such as cameras and steel bars) to protect the contents of their brick and mortar business. While this would be an appropriate measure for a brick and mortar business trying to protect inventory, it would not be appropriate for a business whose inventory is primarily service based or online. We see that risks are not the same for

I. Alsmadi et al., *The NICE Cyber Security Framework*,
https://doi.org/10.1007/978-3-030-41987-5_6

everything; risks are unique and situational and should always be analyzed and dealt with according to how they affect the potentially affected business.

As stated previously during the four-step process, assessing and prioritizing risks are important aspects of the risk management process. Risks are generally determined by identifying the potential risks and multiplying it by the potential vulnerability of the business. Opportunities are multiplied by the known vulnerabilities of a business in order to determine the potential impact on the business at the current state. As we have previously talked about, determining risks is important because it allows for a business to understand the impact that potential risks would have on a business in the event they were to occur, giving an organization the chance to prevent and correct risks accordingly [2].

Although risk management tends to have negative connotations stemming from the word 'risk', there are instances in which risks can be 'positive'. Positive risks are taken when businesses are not entirely sure of an outcome but know if they invest in a certain outcome the potential benefits outweigh the risks. This proactive approach has been visited by organizations such as the Project Management Institute's (PMI) *Project Management Body of Knowledge (PMBOK)* in order to have a more practical approach on project and risk management [3]. Project Management Institute states that the effects of risk management can either be positive or negative. PMI believes that organizations who follow the proactive approach can minimize the negative risks and maximize positive risks.

This chapter addresses the risk management from an insider perspective. Then a detailed case study about risk management on Intelligent Automation in Financial Institutes is presented and discussed.

Risk Management from an Insider's Perspective

Insider risks and threats are a growing concern in the workplace. Companies are doing their best to ensure that intellectual property and unauthorized data never leaves their network. For many, the thought of disgruntle employees with the perfect blend of power, rights or access wreaking havoc on one's network is enough to keep system administrators up at night thinking of multiple solutions to mitigate these scenarios. Yet, as user's rights and roles expand, they acquire new privileges, and some are delegated elevated access because "it's just easier" or "it's what we've always done."

As most know, those are not appropriate answers to this steadily growing problem. In fact, insider threats are one of the most dangerous security threats, and a much more complex issue [4]. Sure, intrusion detection and prevention systems, continuous monitoring and log analysis are all standard tools of the trade to try and catch someone in the act, but they only tell a portion of the tale. Not all insiders operate with nefarious intent. Most admins and power users know that with great power comes great responsibility and do their best to adhere to corporate policies and best practices. However, even the best and most vigilant employees make mistakes because they are human. We live in a world where phishing expeditions are becoming much more believable and sophisticated and breaches more commonplace [5]. Those with the greatest of intentions, knowledge and reputation can be fooled and owned. In other words, no one is safe. These types of users are not necessarily threats, but they are risks.

Insider Data Misuse

Insider risks and threats share a common trait; individuals have legitimate access to an organization's information technology (IT) infrastructure but where they may differ is origin, motivation and action. A risky insider does not become a threat until they act with ill or accidentally harmful intent. Figure 6.1 focuses on the types of privilege misuse and abuse associated with technology according to the 2017 Verizon Data Breach investigations report [6].

Yet, there are many other factors to consider in terms of identifying risky users such as mood (being upset, for example), disgruntled, mental health (depression), family or marital issues (death, dysfunction, separation or divorce), the feeling of entitlement (pay increase, promotion or tenure), etc. All these situations can lead to distraction, poor decision making, a general lack of focus or even worse, incidents of fraud, theft, abuse or sabotage. It is important to follow clues and warning signs to correctly identify these risky insiders, but some risks and threats are not easy to discern as certain individuals wear the mask well.

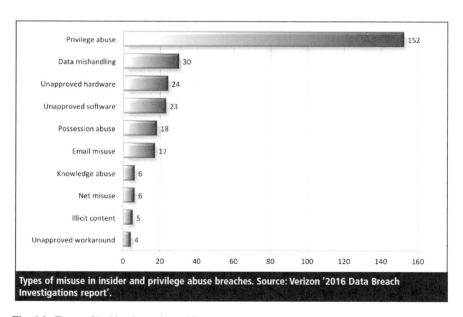

Fig. 6.1 Types of insider data misuse [6]

Identifying these risk and potential threats requires a team with enough freedom, flexibility and finesse to maneuver the potential political and technological roadblocks that lie ahead. Constant communication, transparency and trust amongst teams such as HR, Ethics IT, Security, Legal, and Investigations is paramount [7]. Here's an example of what that may look like: Specifically, the team assigned with gathering then interpreting this data has a pivotal task. The information they provide can make or break someone's career, case and/or investigation. Real data is being used to draw conclusions and make informed decisions whether it comes from a suite of tools, scripts, machine learning or all of them. Usually such team in the enterprise consists from different people with different roles such as Chief Information Officer (CIO), Chief Information Security Officer (CISO), HR, IT Security manager, Incident commander and many others [7].

Certain members of this team will need unfettered access in order to fulfill their tasks properly which requires more privileges on accessing the data. Without this type of access, an organization will never be able to assert their completeness post-incident. In other words, someone must have the complete picture, but what exactly does that mean? Let us use the following example for illustration:

Two employees (John and Jill) are married, work for the same company and are going through an ugly divorce. Jill claims that John has become verbally abusive, may be using drugs and made statements suggesting self-harm. So, Jill goes to work and reports what's happening to a manager who then reports it HR which results in an internal investigation. The investigations team requests data regarding the married couple which includes, email, instant messages, internet history, log on and log off logs, badge access history and social media accounts. The investigations team may use this information to see if John's threatening Jill or anyone else at work, looking up items on the web pertaining to self-harm and using at work because John's manager confirmed that John is taking breaks more frequently. All these factors indicate risk and potential threats to the organization. If John is making threats and using, John may come to the office in an altered state of mind and cause physical harm to Jill, another employee or himself. He may get sloppy and make a mistake when handling a customer's account or do it purposefully because what John's experiencing right now is preventing John from doing the job properly or as some form of revenge. These are all potential warning signs that cannot be ignored. The ethical and moral dilemma is where does the investigation stop? Let us say they work for a bank so, money is involved. Would the business feel comfortable with this team having the ability to view bank account information without question? Does this team need *that* level of access to prevent a risk from becoming a threat or do they involve the feds?

Still, there are other risks to consider, for instance, a system administrator who discover they are going to be terminated. Depending on their privileges, they could exact their revenge proactively and impact the business in a harmful manner [8]. HR may try coordinating future terminations with the investigation team and management to prevent employees from learning about their expiring status but, such an effort would have to be tight. Again, communication, transparency and trust are major factors. No matter the approach, the question remains, how do organizations accurately and ethically identify insider risks and threats before they occur?

Information Security and Protection Systems

At the most basic level, organizations should start with a Security Information and Event Management (SIEM) system. SIEMs are comprised of complex technologies to provide a complete view in an organization's infrastructure. They provide event and log collection, log management, reporting and alerting, correlation, and other useful features [10]. Being able to parse logs for human consumption is a major deal. Personnel must be able to understand what they're looking at in order to make effective decision. They are used for much more than insider activity and they are a good place to start.

Next are Data Loss Prevention (DLP) systems. These are used to identify and monitor data at rest, data in transit and data in use. They also allow organizations to control user's interaction with data based on rules and policies. In an investigation, they are useful for providing "who, what, when and where" information through audit logs and reports. Figure 6.2 shows the workflow for a DLP system.

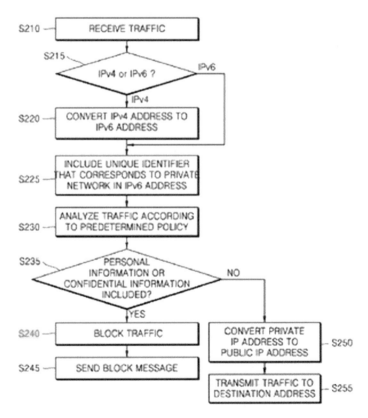

Fig. 6.2 Data loss prevention (DLP) systems workflow

The more modern and desirable approach though, would be using machine learning. Machine learning allows computers to process new conditions through analysis, self-training and observation [9]. This allows the machine to gain experience then use it for decision-making. Tools such as UEBA's (User, Entity and Behavioral Analytics) incorporate machine learning and are gaining popularity in the industry to thwart attacks. UEBA's discover security anomalies by aggregating data via logs and reports, packets, traffic flow, files, user accounts, endpoints and much more. This information is compared to other threat data to help train the model and determine what's normal through repetition thus building a better security. This allows organizations to better understand what their users are doing daily and establish a baseline for network activity [10].

The machine will create its own rules and formulate opinions based what it learns then provide stats and metrics saving personnel from having to do the heavy lifting by performing hours of log sifting [11]. Those are manual processes and history has taught the industry that humans are error prone. Through automation, efficiency is gained, errors are reduced, and repeatability is achieved. Table 6.1 shows how these technologies may be combined to provide a layered defense in an organization. Almost every if not all security course teaches the "defense in depth" methodology and while this chart is not conclusive, it provides an idea of how these tools can be cross-functional in design and application.

An important defense line is the education. It sounds so simple and in theory it is but so many organizations take this for granted. Annual cyber security awareness training typically consists of a couple of videos, slides and/or quiz afterwards. Some places give the *exact* same training every year and do not update the content. That is unacceptable. As the needle moves, education must adapt to sharpen the user community. They need to be aware of the warning signs and they must know that they have a stake in the success of their company. This training should include a

Table 6.1 Layered defense in an organization

Type	Observe	DLP	SIEM	Analysis
Fraud	Social engineering	X		X
	Internal or external collusion	X		
	Network data exfiltration	X	X	X
	Creation of Fraudulent Assets		X	X
Sabotage	Deletion of logs		X	X
	Introduction of unauthorized code		X	
	History of rule violation			X
	Coworker conflict			X
Theft	Solicitation from external parties	X		
	Suspicious travel			X
	Access of outside of need to know	X	X	X
	Physical data exfiltration		X	

section that help users identify internal risks and provide scenarios like the ones mentioned earlier as well as topics such as social engineering, phishing, vishing, fraud, etc.

Case Study: Risk Management on Intelligent Automation in Financial Institutes

Technology has evolved the use of risk management techniques in the financial industry for the past decades and continues to grow. More than 50–60% of the financial industries spend money towards process efficiency improvements and earnings that are more technological advanced to keep up with competitors, keep up with consumer demands, and make things easier and user friendly towards the user [14]. Financial Institutions commonly use intelligent automation tools that are "algorithms based" and learn new trends to identify new risks. Risk management tools like the intelligent automation in the financial industry help identify bad decisions made by a business line, gaps in a process, identify future trends, or even give mass count of impacted loans with the common situation (based on the algorithm criteria). Intelligent automation in financial institutions brings a luxury to companies where there is less human interaction that cannot prevent less errors for the target goal thus meaning providing less risks towards the company. With the improvement in technology in a positive way it can also serve in a negative way in that hackers are becoming more advanced. The number of vulnerabilities and incidents reported in the financial industry has drastically increased over the past decades and will continue to grow [12].

Intelligent automation can be controlled by basic steps that serve across many financial institutes that include identify potential situations that might occur (one off situations), testing, monitoring, and finally controlling. Following the advancement of technology many risk management groups in financial institutes are quickly trying to adjust their thinking and approaches and finding new ways for improvement.

With any great company, you want to hear feedback from the consumer, the very people loans or accounts you are servicing. According to Fig. 6.3 (Security is a major consumer concern with Some Issues About Risk Management for E-Banking) the data suggested with UK online consumers are concerned with computer viruses, credit card number theft, interception of personal date to just name a top few. These financial institutes takes these polls and various feedbacks into consideration and use these as foundation base platforms for risk management.

The financial institutes should keep their knowledge up to date in terms of risk management from an electronic banking view not only in the United States, but across the world. Moreover, these institutes should keep up with regulatory requirements and need to make sure they are in the best interest of the customer to avoid negative reputation. Risk Management in intelligent automation helps all those needs [13].

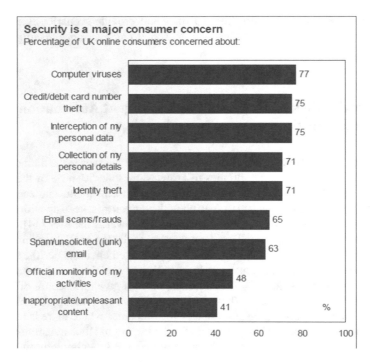

Fig. 6.3 Percentage of online consumers concerned about data security in UK

Current Solutions and Designs

In financial industries –similar to other sectors - every company wants to make sure that business lines are running normal with the least amount of errors (or none). That's why electronic risk management is important to the finance industry and when you're having to manage federal regulations, etc. It is important for these businesses to weigh out the risks involved for any kind of changes and updates you are doing for the company. "The speed of change relating to technological and customer service innovation in e-banking is unprecedented. Historically, new banking applications were implemented over relatively long periods of time and only after in-depth testing. Today, however, banks are experiencing competitive pressure to roll out new business applications in very compressed time frames—often only a few months from concept to production" based upon (Risk management for internet banking). The game plan for most of these financial industries is constantly be aware of all risks and regulations that are being implemented and quickly adapt to changes. This involves continuous meetings and utilizing all resources and expertise in the matter.

Board members and management over sight is important and requires constant engagement with them as they play one of the key roles in whom you are trying to satisfy in achieving your goal. Typically in most financial industries they also hold control of the risks associated to the activities and will at times hold sole responsibility to the direction based on the information provided in managing the risks. Board members and management should be aware of all the factors before making an important decision in weighing in the pros and cons of the risks involved, especially when it comes to intelligent automation. Accountability does come into play when separating the roles in who is in charge of what process when management is reviewing all the information to make a decision, especially when dealing with banking, mortgages and customers payment. Many intelligent automation tools are ran to make sure customer payments will be paid off in time, customer's current interest rate and payment are currently being displayed in the computer systems. Letters to customers are also playing a role in the intelligent automation where automated letters are generated and sent out to the customers given the data placed in specific fields in the banking programs. All these intelligent automation programs are great improvements to financial industries but can also be a problem when something isn't being ran efficiently and an error is identified. That information must be presented to a senior level management for review. When running through the risks many financial institutes commonly go through a strategic a planning in a cost/reward analysis. That consists of making sure they have the appropriate groups and teams that can provide certain expertise for every situation given factors of foreclosure, bankruptcy, certain types of products being offered accordance to terms of the mortgage note. Management typically set a foundation of risk assessment in roles and asks reasons in how the error occurred? what we can do to fix the issue?, determine if other customers were impact, and determine if there could be future impacts. All this information can be pulled through system automation reporting in a high-level algorithm where data is retrieved in from the computer programs that houses the data for each customer's loan information, and scrubs the loans based on the criteria being asked.

With any risk management there has to be security controls in place to prevent any kind of tampering and alterations to a process. Some security controls include authentication of users whom are working on the process and have specific passwords, challenge and response systems, and public key infrastructure related to the project [14]. Specific individuals are granted access to specific portals within the company that can hold confidential information on project. Monthly report can be running in determining the last time the user accessed the portal and send out emails to manager/supervisors to determine if the user still needs access to these certain portal due to the user not using the site/portal for a long and extended time period. Working in the financial industry many technology security controls are in place, due to information confidentiality is level is high. Some classified information to the customers includes customers loan information like home address, account numbers, telephone number, social security number, pay check information, and bank statements. There are also customers whom are in the witness protection program where it is the financial institutes responsibility to have certain controls in place to protect these customers identity and personal information.

It is important that financial institutes fully authenticate employees to make sure the correct person is viewing the information and the specific individuals are able to view data information. All these security controls help avoid the risk of being compromised and have a breach in data.

Not only are there internal security controls in place but there has to be external security controls that help prevent external attacks on valuable information from either outside source looking to breach data in large amounts. Capital One just recently had a situation where their information was breached and an external user had taken a large amount of personal data. This had also happened with other major companies like Facebook a couple of years back. This can ruin the reputation to Capital One and Facebook as it loses trust to the customer and takes a long time to gain back the trust from the customer. As can be seen from Fig. 6.4 (Increasing number of data breaches by Market Watch), data breaches have skyrocketed over the past years and in the financial industry. Individual attacks can also play into effect where someone can claim they are a customer when they are really not. They try to key information from either bank mobile devices or banking online services in an attempt of falsely representing the customer [15].

Certain security questions are always asked when users sign on, to make sure they are talking to the customer and certain key logins are always asked to prevent any cyber bots trying to gain access. Employees are also trained not to click any links via email of unknown sources and are not coming from a credited source. Monthly training and tests are commonly ran for employees in where any suspicious emails should be reported for spam reviews all in attempt to protect any kind of data breaches since the employees have access to classified information [16].

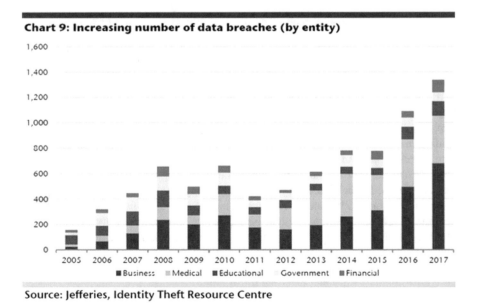

Fig. 6.4 The number of data breaches grouped by entity

Separation of duties also play an important part in data security control for intelligent automation. It is important to separate the duties so one person cannot hold true control of a project to avoid fraud by any individual. It is important that every piece has outlined duties that can be second level reviewed by sources within the project or groups that serve as a quality control. Quality Control Specialists are designed specifically going behind a project and serving every job function is being performed correctly based on the instructed procedures and also can provide random quality checks.

Finally, financial institutes "strengthen information security controls to preserve the confidentiality and integrity of customer data. Firewalls, ethical hacking tests, physical and logical access controls are some of the methods available" per E-banking: risk management practices of the Estonian banks. Risk Management on Intelligent Automation also consists on legal and reputation issues that go on within the financial institute. They can be divided by three following factors like privacy, availability and incident response.

Challenges and Problems

There are many challenges faced in financial institutes when dealing with intelligent automation. Given the exposer to new information technologies most often, it is very difficult to keep up with the pace of adjusting line business to have the most up to date technologies and protecting your information. Not only that, but have the employees that will be able to easily adapt to upcoming changes. In the financial industry, for intelligent automation, run into factors every single day for the roles as lines of business are identifying new gaps and holes. For example, in the mortgage side a lot of payment calculations for Adjustable Rate Mortgages are automated due to the amount of loans being serviced at larger companies. There is no way individuals can calculate a new payment for all the customers given enough time notification deadline, and the number or portfolio of loans being serviced. Some portfolios of one kind of product being serviced can be over 500,000 to 750,000 loans alone. Financial Institutes rely on intelligent automation to correctly calculate the new payment and interest to the customer given the data information set in certain field in the banking database housed for that customer. Having to rapidly keep up to new changes and regulatory requirements can bring risks. It is important when having a game plan of running through all the risks involved when adjust to new changes and making sure the employees are fully aware of situation. So that they can monitor and raise their hands if they see any or issues during the transition in order to come up with a game plan to fix those situations [17].

Financial institutes "applications are typically integrated as much as possible with legacy computer systems to allow more straight-through processing of electronic transactions. Such straight-through automated processing reduces opportunities for human error and fraud inherent in manual processes, but it also increases dependence on sound systems design and architecture as well as system interoperability and operational scalability" per (A comparative analysis of current credit risk models). This especially poses a problem in larger financial institutes in where certain programs that were built years back to house particular loans and can only view certain loans in one program. Other mortgages and loan products were particularly housed in another program called MSP (Mortgage Secured Platform). The problem with the home equity loans be serviced in a different program, SHAW, is that the program was being outdated and not everyone had access to SHAW and people were become less familiar in how to work the program, because of the program being outdated and incapable to manage the home equity loans. A giant project took into effect where they wanted to move the home equity loans over to the newer program of MSP to house all loans in one centralized location. The complication of this project became more troubling then expected as the program, MSP, can only house certain fields and could not quite capture all the important fields a home equity performs and a lot of home equity loans calculate daily simple interest (DSI). Meaning, these loans calculate interest based on when the customer makes a payment, not a standardized monthly interest format like a typical mortgage is calculated in programs in MSP. This project had to bring in new innovation from tenure people, whom worked on home equity loans, people work on the MSP programs daily, and programmers to build new fields and criteria in MSP to where they can identify these loans and make sure they are calculating the correct payment. The process of moving over the loans consisted of testing out certain fields, identifying the risks, finding a control on issues and possible issues in the future that took a six-month process. With the luxury of intelligent automation, the tools were able to successfully shift the loans over and continue to monitor for any kind of unknown glitches and errors. Continuing with risk aspect, there is staff in place to perform random check samples in making sure payments are correctly calculated to avoid any kind of impacts from the customer and also avoid any kind of fines.

Summary

Figure 6.5 depicts the rising cost of not protecting against malicious insiders and there are no signs of the costs losing momentum. This is not meant to serve as a definitive guide to insider risk and threat mitigation. The goal was to gain an understanding of what may cause an employee to be go from being a risk to a threat while educating and providing insight on where an organization may begin to detect them. Insider risk managementt and threat programs are growing across the industry. Businesses spent much of their time and focus preventing malicious actors from breaching their data that many did not consider the cost of letting one from the inside escape with their data. In 2019, the chances of accepting that kind of risk are much less than they were even going 2 to 3 years back [18]. With new rules such as GDPR and governments levying heftier fines for breaches, organizations are doing what it takes to "batten down the hatches."

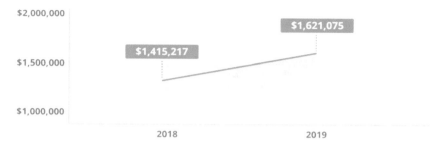

The average cost of insider attacks keeps rising*

* Data provided by Accenture & Ponemon's 2019 Cost of Cybercrime Study

Fig. 6.5 The average cost cybercrimes caused by inner attacks

The case study about risk management in intelligent automation in financial institutes emphasized that the financial institutes has to stay tuned and up to date with efficient risk management techniques specially with the huge increase in cyberattacks on these institutes. Moreover, the concept of risk management is very broad and it is hard to single out this tiny scope of process because they financial institutes applications work with other risks in other lines of business in order for each end to work successfully. It is true that risk management can be all kinds of variables from the technology side but also on the business end. The future of risk management will involve propose unique models where all financial institutes will adopt and make one uniform model.

References

1. Kim, D., Solomon, M. (2018). *Fundamentals of information systems security. Information systems security & assurance series*. https://online.vitalsource.com/#/books/9781284128567/cfi/6/26!/4/2/6/2/32/6/2@0:56.5.
2. Jararweh, Y., Al-Sharqawi. O., Abdulla, N., Lo'ai, T., & Mohammad, A. (2014). High-throughput encryption for cloud computing storage system. *International Journal of Cloud Applications and Computing (IJCAC)*, 4(2), 1–14.
3. Agrafiotis, I., Eggenschwiler, J., Nurse, J. R. C. (2016). Insider threat response and recovery strategies in financial services firms. *Computer Fraud & Security*, 2016(11), 12–19. https://doi.org/10.1016/S1361-3723(16)30091-4.
4. Schultz, E. E. (2002) A framework for understanding and predicting insider attacks. *Computers & Security*, 21(6), 526–531. https://www.sciencedirect.com/science/article/pii/S016740480201009.
5. Al-Haija, Q. A. (2019). Autoregressive modeling and prediction of annual worldwide cyber-crimes for cloud environments. In *2019 10th International Conference on Information and Communication Systems (ICICS)* (pp. 47–51). IEEE.

6. Verizon (2017) *Data breach digest: Perspective is reality.* https://enterprise.verizon.com/resources/reports/data-breach-digest-2017-perspective-is-reality.pdf.
7. Ambre, A., Narendra, S. (2015). Insider threat detection using log analysis and event correlation. https://doi.org/10.1016/j.procs.2015.03.175.
8. Verizon (2019) *Data breach investigations report.* https://enterprise.verizon.com/resources/reports/dbir/.
9. Sarkar, K. R. (2010). Assessing insider threats to information security using technical, behavioural and organizational measures. *Information Security Technical Report, 15*(2010), 112–133. https://doi.org/10.1016/j.istr.2010.11.002.
10. Miller, D. (2011). *Security information and event management (SIEM) implementation.* McGraw-Hill.
11. Maher, D. (2017) Can artificial intelligence help in the war on cybercrime? *Computer Fraud & Security 2017*(8), 7–9. https://doi.org/10.1016/S1361-3723(17)30069-6.
12. Shashanka, M., Shen, M.-Y., Wang, J. (2016). User and entity behavior analytics for enterprise security. In *2016 IEEE International Conference on Big Data (Big Data)* (pp. 1867–1874). IEEE.
13. Mayhew, M., Atighetchi, M., Adler, A., & Greenstadt, R. (2015). Use of machine learning in big data analytics for insider threat detection. In *MILCOM 2015–2015 IEEE Military Communications Conference* (pp. 915–922). IEEE.
14. Härle, P., Havas, A., Kremer, A., Rona, D., & Samandari, H. (2016). The future of bank risk management. *McKinsey & Company.*
15. Möckel, C., & Abdallah, A. E. (2010). Threat modeling approaches and tools for securing architectural designs of an e-banking application. In *2010 Sixth International Conference on Information Assurance and Security*, (pp. 149–154). IEEE.
16. Harris, E., & Younggren, J. N. (2011). Risk management in the digital world. *Professional Psychology: Research and Practice, 42*(6), 412.
17. Lo'ai, A. T., & Saldamli, G. (2019). Reconsidering big data security and privacy in cloud and mobile cloud systems. *Journal of King Saud University-Computer and Information Sciences.*
18. Spooner, D., Silowash, G., Costa, D., & Albrethsen, M. (2018). *Navigating the insider threat tool landscape: Low cost technical solutions to jump start an insider threat program 2018 IEEE security and privacy workshops (SPW)* (pp. 247–257). San Francisco, CA. https://ieeexplore.ieee.org/abstract/document/8424656.

Chapter 7
Software Management

Izzat Alsmadi

K0009: Knowledge of Application Vulnerabilities

Vulnerabilities General Statistics

Between 2015 and 2018, the vulnerability count had spiked for most vendors, 2016 is very significant. This was cross-referenced to the Heartbleed incident from a little over two years ago. There were 36,536 total vulnerabilities from all ten vendors, with an average of 3653.6 per vendor. Microsoft, of the ten vendors, had the highest vulnerability count of 6031 while Mozilla had the lowest count of 2048. Oracle had the second-highest vulnerability count of 5506, while Redhat had the second-lowest vulnerability count of 2147. Apple had the third-highest vulnerability count of 4273, and Linux had the third-lowest vulnerability count of 2185. IBM had the fourth-highest vulnerability count of 4194, while Adobe had the fourth-lowest vulnerability count of 2798. Google had the fifth-highest vulnerability count of 3726, while Cisco had the fifth-lowest vulnerability count of 3628. Microsoft made 16.5% of the total vulnerabilities, Oracle made up 15.1%, Apple made up 11.7%, IBM made up 11.5%, Google made up 10.2%, Cisco made up 9.9%, Adobe made up 7.7%, Linux made up 6.0%, Redhat made up 5.9%, and Mozilla made a meager 5.6%.

Under the bypass category, Adobe had 204, Apple had 398, Cisco had 301, Google had 286, IBM had 238, Linux had 111, Microsoft had 434, Mozilla had 202, Oracle had 82, and Redhat had 132. 10,138 vulnerabilities of this kind were identified. For code injection vulnerabilities, Adobe had 1944, Apple had 2133, Cisco had 589, Google had 533, IBM had 525, Linux had 245, Microsoft had 2839, Mozilla had 762, Oracle had 206, and Redhat had 362. 10,138 in total were brought to attention.

With CSRF, Adobe had 10, Apple had 4, Cisco had 109, Google had 4, IBM had 128, Linux had 0, Microsoft had 3, Mozilla had 21, Oracle had 2, and Redhat had 22. 303 vulnerabilities of this category were revealed. For directory traversal,

© The Editor(s) (if applicable) and The Author(s), under exclusive license to Springer Nature Switzerland AG 2020
I. Alsmadi et al., *The NICE Cyber Security Framework*,
https://doi.org/10.1007/978-3-030-41987-5_7

Adobe had 3, Apple had 25, Cisco had 68, Google had 11, IBM had 78, Linux had 0, Microsoft had 22, Mozilla had 12, Oracle had 30, and Redhat had 25. 274 total directory traversal vulnerabilities were discovered.

In the list of DoS vulnerabilities, Adobe had 868, Apple had 2187, Cisco had 1441, Google had 1209, IBM had 519, Linux had 1197, Microsoft had 1518, Mozilla had 645, Oracle had 420, and Redhat had 573. This made a whopping 10,577 total DoS vulnerabilities. On file inclusions, Adobe had 1, and Cisco and IBM had 2, making a total of 5 file inclusions.

Under the Gain Information category, Adobe had 139, Apple had 137, Cisco had 45, Google had 73, IBM had 72, Linux had 58, Microsoft had 118, Mozilla had 57, Oracle had 14, and Redhat had 38. This made 751 total vulnerabilities of this type. With Http Request Splitting, Adobe had 3, Apple had 2, Cisco had 7, IBM had 23, Linux had 0, Microsoft had 2, Mozilla had 3, Oracle had 3, and Redhat had 5. This, alone made only 48.

For Memory Corruption, Adobe had 914, Apple had 1317, Cisco had 27, Google had 221, IBM had 10, Linux had 124, Microsoft had 1483, Mozilla had 390, Oracle had 32, and Redhat had 69. Memory corruption made 4587 total vulnerabilities. In Overflow, Adobe had 1186, Apple had 1660, Cisco had 281, Google had 1661, IBM had 400, Linux had 349, Microsoft had 1753, Mozilla had 397, Oracle had 189, and Redhat had 326. In total, there were 8202.

In Privilege Elevation, Adobe had 39, Apple had 489, Cisco had 397, Google had 563, IBM had 852, Linux had 400, Microsoft had 1000, Mozilla had 175, Oracle had 202, and Redhat had 269. This totaled to 4386. For SQL injection, Adobe had 1; Cisco had 15, Google had 2, IBM had 29, Linux had 0, Microsoft had 1, Mozilla had 5, Oracle had 39 Redhat had 1. Only 93 SQL injections made the vulnerabilities list. For XSS, Adobe had 104, Apple had 113, Cisco had 290, Google had 74, IBM had 937, Linux had 0, Microsoft had 210, Mozilla had 146, Oracle had 49, and Redhat had 86. 2009 cross-site scripting vulnerabilities were in the top-ten list. In the total of all the vulnerabilities in the list, there were 43,761. As an example of one category of vulnerabilities, Fig. 7.1 shows the distribution of code injection vulnerabilities across major US technology vendors.

Vulnerabilities with DBMSs

In 2015 the Oracle Corporation was the second-largest software maker, in terms of revenue, after Microsoft. They provide OS and database applications, cloud engineering systems, enterprise software products and provide extensive support for their entire suite (News, tips, partners, and perspectives for the Oracle Solaris OS, Feb 2013).

Along with the popularity, accolades, and income that comes with a top-ranking and massive customer base, there also exist many pitfalls. A large number of users will attract a large number of attackers, seeking to exploit the vulnerabilities of these

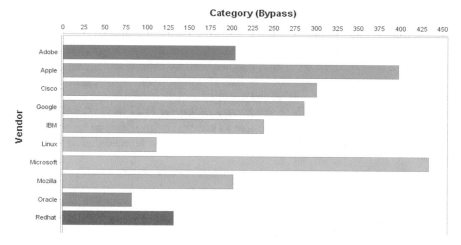

Fig. 7.1 Major vendors and code injection vulnerabilities (*Source* CVE and NVD websites)

applications to gain access to the sensitive information belonging to both the end-users of these applications, as well the companies who maintain the applications. Consumers need to be able to process and distill this information in order to decide which products are best for them. Administrators need to keep abreast of new threats as they manifest to adjust and protect their systems accordingly. Repositories such as CVE Details, which gets its data from the NVD, a security vulnerability database, keep track of this information and offer it online.

Users can obtain this data from NVD for analysis in order to help with this decision. For the sake of this paper, we will use data retrieved from NVD using a custom java program which will scan the JSON files, and convert the findings to a .csv format. With this data, we hope to be able to ascertain a connection between the number of vulnerabilities and the usage, or market share, of a given product.

A cursory glance at the Solaris OS results indicates a marked difference between Sun and Oracle. In 2015 the Solaris 9 reached end-of-life, with Oracle no longer providing support, leaving Solaris 10 and 11 as the current OS options (News, Tips, Partners, and Perspectives, for the Oracle os, Feb 2013). The organization saw a spike in vulnerability issues with 163, up from 65 the year before, perhaps due to the drop off of Solaris 9, the last version of Solaris fully created and maintained by Sun.

Solaris is a type of Unix operating system, originally developed by Sun Microsystems in the late 1980s as a merger of the most popular Unix variants on the market at the time: Berkeley Software Distribution (BSD), UNIX System V, and Xenix. It was initially known as System V Release 4 (SVR4) but was changed to Solaris 2 starting with the next release, with the previous version of the Sun OS retroactively renamed Solaris 1. Sun continued to develop and maintain Solaris until purchased by Oracle in 2010 (News, Tips, Partners, and Perspectives, for the Oracle os, Feb 2013). In 2005, Sun released most of the codebase and founded the OpenSolaris open-source project to encourage a developer and user community around their software. The main difference between Solaris and OpenSolaris is the level of support provided by the Sun. OpenSolaris received little to no support beyond that provided for free by the community of developers. Solaris support had to be contracted by a consumer. In 2010, Sun was acquired by Oracle, and OpenSolaris was discontinued. Effective March 2010, Solaris 10 was no longer available for free (support came with a cost) and was placed under a restrictive license and allowing users only a 90-day trial period. Upon the release of Solaris 11 in 2011, Oracle allowed Solaris 10 and 11 to be downloaded free of charge and available to be used indefinitely with a support contract, so long as it is not for a commercial or consumer use.

Certain trends emerged when analyzing the NVD vulnerability listings for Solaris and OpenSolaris. Many of the vulnerabilities for OpenSolaris were not assigned to a CWE category, and none of the CVE Id's for Oracle OpenSolaris had a corresponding CWE category, with all of them being listed as NVD-CWE-no info. This was probably due to the lack of official support by Sun and Oracle, who eventually discontinued the OS in 2010.

There are hardly any CWE designations attributed to Solaris in the early years, while it was still being developed by Sun. The vast majority of CVE's have a designation of "NVD-CWE-Other," indicating that the vulnerability either predated the use of CWE's or that NVD, which uses only a subset of CWE's provided by CVE Details, had a designation that was not used by NVD. Researching the summaries of these new vulnerabilities suggests that they should have a CWE id of 119, as many of them involved buffer overflow issues. Updating this list would make buffer issues the number one vulnerability category for Sun Solaris.

After the transition, there is a significant increase in reported CVE entries, yet they were much less severe. With Sun in charge of development, almost 60% of all CVE's had a score of 7 or higher. Once Oracle was entirely in charge of support, particularly after the development of Solaris 11, the first version of Solaris entirely under Oracle's umbrella, the severity of CVE entries seemed to stabilize, with fewer than 19% of entries having a score higher than 7.

Database management systems seem to show a pattern of vulnerabilities that parallels market share. The more popular the database, the more vulnerabilities reported. Oracle, having the highest market share, led the way with well over 500 vulnerabilities listed (DB-Engines Ranking, Dec 2018). MySql was a distant second in terms of market share and had a corresponding number of vulnerabilities with over 240. Microsoft SqlServer seems to buck the trend, having the third-largest market share, but only 86 vulnerabilities listed in NVD, while Postgresql, in fourth place, has over 100. However, a closer look at SqlServer's CVE entries shows that over 25% of the reported CVE entries have a score of 8 or higher (OWASP JSEC CVE Details, Sept 2016). The high percentage of higher scores indicates that the number of entries may be underreported. Except for SqlServer, the trend seems to indicate a correlation between market share ranking and the number of CVE entries.

K0039: Knowledge of Cybersecurity and Privacy Principles and Methods that Apply to Software Development

Vulnerabilities in software applications can be traced back to different types of sources. Those include the programming language in which the program is developed in, the program itself and how software developers to design and construct the code, the platform or environment in which the software is deployed in and also the users who are going to use the software. In this section, our focus is only on the first part of the vulnerabilities that are related to the programming language or that the programming language can have an impact.

Buffer and Stack Overflow

Applications are loaded to memory when they are under usage/execution. A buffer overflow exists when an application tries to write data beyond the buffer region. A buffer is a reserved memory space for a program variable. Buffer overflows can occur from programs' errors, but they can also be crafted intentionally by hackers. As a result, an application can crash attackers may gain system access without proper authorization or may achieve a privilege elevation. Applications' memories are stored in either the application stack or the general heap storage.

Overflow attacks occur when attackers manipulate data input to a program. Such data manipulation causes problems to temporary data stored in memory. In overflow attacks, the attacker tries to write data beyond the location that software designers intended for the input data. A memory stack is a memory location that is only accessible from the top. It is used to store local variables and function addresses. In a stack, a push operation inserts new data to the top of the stack, and a pop operation removes data from the top of the stack.

Ultimately, writing secure software code (e.g., with only safe constructs) can ensure avoiding overflow attacks. For overflow attacks, in particular, programmers should have code to check (not only assume) that data values entered by users will not go beyond the limit decided by the program.

Programming languages are different in their interaction/response to overflow attacks. Some programming languages enforce the use of only safe programming constructs.

Consequences of overflow attacks can be either to cause code, data corruption or program failures generally. Alternatively, in some cases, a successful overflow attack may make attackers succeed in creating a program or system backdoors or execute their malicious codes.

Memory Leak and Violation Issues

When users launch a program, the program will be loaded from the disk storage to the memory. Different programs can consume different memory sizes. Once users close those programs, they will be cleaned from memory to enable other applications to reclaim their memory and other allocated resources. This ensures better system performance and memory utilization. One of the main features in computer hardware that is continuously increasing is memory size. As a consequence, users may not notice memory leak issues as previously unless if such a leak is significant.

Different mechanisms exist in operating systems and programming languages to clean up memory and de-allocate memory resources that are not needed anymore. A memory leak occurs as a failure to do that where mechanisms to de-allocate memory resources have some problems. In some cases, those different mechanisms may ignore memory cleaning as they assume that other mechanisms will do that.

Attackers, knowing that an application may have a memory leak, will try to send frequent requests to the application. As the application lacks a proper method to clean memory after each service call, this may ultimately cause a memory leak. As a result, the application may deny further legitimate users' services. While this problem is classical and has been in the security.

K0040: Knowledge of Vulnerability Information Dissemination Sources (e.g., Alerts, Advisories, Errata, and Bulletins)

Websites such as NVD (https://nvd.nist.gov) and CVE (https://www.cvedetails.com/) continuously record security vulnerabilities based on a US national standard since the early 90s.

The Common Vulnerability Scoring System(CVSS) provides a quantitative mechanism to reference information vulnerabilities. Vulnerability defined by the Common Vulnerabilities and Exposures (CVE) as: "A weakness in the computational logic (e.g., code) found in software and hardware components" (https://nvd.nist.gov/vuln). The CVSS' numerical score reflects the severity of exploits in a qualitative representation such as low, medium, high, and critical to helping prioritize the vulnerability management process. This Common Weakness Enumeration (CWE) classifications are inconsistent and inclusive in more than one of CWE's categories. The CVSS categorization focuses on implantation issues such as XSS or SQL but overlooks the design and architecture of the software.

For example, the impact of XSS flow receives "low" weight in the meantime; there is no input validation mechanism; in this case, the given weight can be elevated to medium or high. The priority of design flows should be considered in a higher category than implementation or use separate factors in categorization. The business impact and the use of products are different. A Safety and reliability product's scores should be evaluated based on business impact. The value of a score should be dynamic; the organizations should reevaluate the sore since the likelihood of a weakness frequently changes due to improved detection techniques and technology advancement. The frequency of these changes could negatively impact the remediation process of a weakness whose category level has been reduced (https://cwe.mitre.org/cwss/cwss_v1.0.1.html).

Examples of Vulnerability in Industrial Systems and IoT

While the technology and applications for the Internet of Things (IoT) devices are not new, easy to use, and affordable implementations were not available to the home user. Rather, these devices were designed and remained prevalent in manufacturing and industrial settings. In these settings, IoT technology was tested under demanding conditions and continuously improved. In the last few years, however, the consumer market for IoT devices has exploded. Consumers now have a vast array of smart technology to monitor and automate their buildings and homes. This massive influx of devices may present a problem. Industrial grade IoT devices, like most technology, have inherited vulnerabilities. Since they are internet-facing devices, they are susceptible to internet-based attacks. Have these vulnerabilities found their way to the consumer market? This work explores the vulnerability trends for consumer-grade IoT devices by collecting IoT data from shodan.io as well as vulnerability data from the NIST NVD database.

EtherCAT (Ethernet for Control Automation Technology) protocol was standardized as IEC 6115817, within the standardization of Fieldbus to fit Ethernet into the field environment. EtherCAT was designed to minimize packet processing time to be suitable for real-time applications.

EtherCAT was derived from Ethernet technology, and hence it suffers from all security vulnerabilities that regular Ethernet networks would suffer from.

DNP3 (Distributed Network Protocol) operates at the data link, application, and transport layer levels. DNP3 has advanced features that make it robust and highly available, but it was not made to be immune against determined hacker's attacks.

At the data link layer, DNP3 uses Cyclical redundancy check (CRC) calculation for error detection. However, CRC does not add that much to security since attackers can manipulate CRC sum value easily.

At the application layer, some efforts were made to add more Secure Authentication features. For example, newer standard Std. 1815–2012 enables the use of public/private keys in addition to pre-shared keys. Vendor errors due to the complexity of the protocol lead to some security vulnerabilities.

Table 7.1 shows examples of top industrial systems' vulnerabilities with their CVE and CWE IDs.

Table 7.1 Top industrial vulnerabilities with their CWE IDs and CVSS scores

CVE-2017-7679 (119-7.5)	CVE-2012-0814 (255-3.5)	CVE-2016-8612 (20-3.3)	CVE-2017-15906 (275-5.0)
CVE-2014-0231 (399-5.0)	CVE-2014-0098 (20-5.0)	CVE-2010-2068 (200-5.0)	CVE-2013-6438 (20-5.0)
CVE-2017-15906 (275-5.0)	CVE-2017-7668 (20-7.5)	CVE-2017-3167 (287-7.5)	CVE-2017-3169 (476-7.5)
CVE-2013-1896 (264-4.3)	CVE-2010-4755 (399-4.0)	CVE-2010-4478 (287-7.5)	CVE-2010-5107 (NA-5.0)
CVE-2016-10708 (476-5.0)	CVE-2011-4327 (200-2.1)	CVE-2014-1692 (119-7.5)	CVE-2011-5000 (189-3.5)

Some observations based on Table 7.1:

- Top industrial systems vulnerabilities are related to software applications, particularly: Apache, and OpenSSH.

While some (Common Weakness Enumeration) CWE categories such as (20, 119, 287, 476, 287) frequently appear in industrial systems' vulnerabilities. Most of those vulnerabilities can fall under Poor Network Protocol Implementations (e.g., lack of input validation, weak authentication, etc.). ICS-CERT 2016 reported the following 4 CWEs as the first categories (CWE-121: Stack-based Buffer Overflow; CWE-20: Improper Input Validation; CWE-79: Cross-site Scripting; and CWE-122: Heap-based Buffer Overflow.

- Additionally, vulnerabilities can fall in many other categories as well. ICS-CERT 2016 reports show that the majority of the industrial related vulnerabilities have CVSS values above 5.

CVSS scores recorded for those vulnerabilities also vary in the scale from (0 or 1 to 10). Highest in terms of counts is the CVE-2017-7679, CVSS score 7.5, CWE ID: 119, Overflow: In Apache httpd 2.2.x before 2.2.33 and 2.4.x before 2.4.26, mod_mime can read one byte past the end of a buffer when sending a malicious Content-Type response header.

K0079: Knowledge of Software Debugging Principles

Software debugging involves activities accomplished by software developers in response to feedback about errors or failures.

Software modules or components are developed by programmers or software engineers. Software testing is usually between programmers who perform unit or component testing. Other types of testing (e.g., functional, integration, black-box testing, etc.) are accomplished by internal or external testers.

Reports of software testing include different types of categories of problems or failures in testing software components. Those errors or failures trigger a new cycle by software programmers to debug the code and find the symptoms then causes of errors or failures. In the next stage, developers find and fix the causes of those problems. The effort of software developers in debugging stage can be divided under the following sub-tasks:

- Start by evaluating the symptoms of errors or problems. Developers should be able to recreate errors based on how they are described by software testers or users who reported the error. Error reporting should cover enough details to help to recreate the error. Developers may reject or return errors if they could not recreate it. They may also reject them if they represent extra/additional features rather than fixing problems with existing code.
- If developers can recreate the errors described by software testers or users, they can mark those errors or bugs as "work in progress" or "open" and start their diagnosis analysis to find the roots of those symptoms. Their work is very similar to the role of medical physicians who test their patients to find the causes/diseased based on observed symptoms. In software projects and based on the size of the code of the project, roots of errors may or may not be in the same method/class where errors are exposed to users. The causes of errors can be distant from symptoms. They can be in other classes or packages in the same project. They may also come from external libraries or applications.

In some cases, errors can be intermittent, may appear some times and disappear in others, may appear on particular transient scenarios. Errors may be known but ignored as they occur in rare cases or as they have no clear solutions.

- Impact analysis: Once causes of errors are discovered, developers should conduct an assessment to evaluate the impact of changing the code in response to debug fixes. In some situations, bug fixes may cause the rise of other bugs in other areas or locations. Software codes and compilers are designed in ways that they may not reveal all errors at once. In other words, fixing a particular problem may trigger others to arise as they were, hidden, behind this earlier one.

In the scope of errors and debugging, different terms are used: bugs, errors, problems, faults, and failures. Some references distinguish those terms to indicate low level or high/user level problems while in other references, those different terms are used interchangeably.

K0080: Knowledge of Software Design Tools, Methods, and Techniques
K0081: Knowledge of Software Development Models (e.g., Waterfall Model, Spiral Model)
K0082: Knowledge of Software Engineering

The software process is a set of complex activities required to develop software systems. There are many known software process models such as waterfall, spiral, incremental, Rational Unified Process (RUP) and agile models. All those process

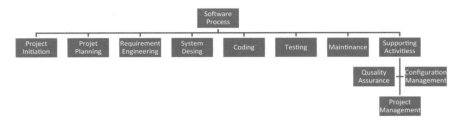

Fig. 7.2 Software process activities [1]

models carry out the same mandatory activities, e.g., requirements elicitation, software design, coding, testing, evolution, and project management. However, they are different from each other based on the way they perform such tasks. For example, a waterfall model follows a straightforward process that starts with the requirement. Once the requirement stage is finished, the design stage is started and so on until finishing all software activities. Such a traditional model does not allow backlinks from a stage to the previous stages. Furthermore, development stages are distinctly divided and separated, and it does not allow overlapping of stages. On the other hand, most of the other development methodologies are flexible in this manner and allow incremental cyclic process wherein each cycle a small part of the project is developed, and all process activities are executed cyclically.

Figure 7.2 describes the typical software process activities. Project initiation and planning are the first tasks in a software development lifecycle, where the project boundaries and feasibilities are generated. After the approval of customers and developers, detailed level planning is carried out by the development team, focusing on the estimation of cost, resources, and scheduling. In the next phase system and user, requirements are gathered, documented, prioritized, and finalized. The primary deliverable of this initial stage is software requirement specification (SRS), which highlights system and user requirements. In this initial stage, the problem domain is analyzed looking for why we are developing this software, who asked for this solution, what were the problems in the domain or the existing system (if a system exists). Software and system analysts are the people who usually perform activities in this task or stage. They meet with domain users and stakeholders in trying to analyze the current system and specifying requirements for the sought solution.

In the next stage, a design is proposed (i.e., documents, graphs, pseudo code, etc.) that fulfills all requirements in the SRS. Traceability (i.e., explicit link between SRS and design) should be examined to make sure that all requirements specified in the SRS are handled and at the same time, no unnecessary elements exist in the design. After the design, the implementation stage starts to translate the theoretical solution from design into code. Testing occurs in each stage to make sure that every deliverable is developed as expected. Nonetheless, there is a separate testing stage after coding to test the significant deliverable of the software project (i.e., the code or the program). Once the software is handed over to the customer, it may also require further changes; this phase is termed as the maintenance stage. Along with these core activities, different supporting activities are carried out in the software development process. These include quality assurance, configuration management, and project management activities.

There are several factors that affect the decision of selecting the right process model. This includes the flexibility/stability of the requirements, nature of the software project, time to market, amount of testing, documentation, and quality required, the level of competition and several other factors that can be related to the software, the users, the domain or the environment.

The remaining of the chapter is structured as follows. Section 2 describes software process models in the context of the subject of the chapter, and Section 3 describes some of the related work. Section 4 introduces open-source projects and e-coding projects and websites in the scope of software development. Section 5 introduces the elements that should be modified in traditional models to fit the view of open-source and e-coding projects and is followed by a conclusion.

Software Process and Project Models

In a previous paper, we described teaching software engineering based on four concerns: software processes, projects, people, and products [2]. We will use a similar approach to describe software project models.

Software Process-Oriented Models

Most software development models described in software engineering literature are process-oriented. This means that each model is specified and distinguished from other models based on the way activities are performed. In one classification, process models can be divided into forwarding process models such as waterfall model that have all process activities in a straightforward manner in comparison with cyclic process models such as agile, incremental, RUP, etc. [2] which advocate multiple cycles of developmental activities.

Every software model is a process model, which means that they focus only on one element of the software project and ignore the impact of the other factors on the software project. For example, based on the nature of the software project, an agile development process for building safety, critical software can be different from an agile process for enhancing existing Human Resource Management (HRM) software. We will focus on some of the aspects that can be used to distinguish models based on the project, people, or product types.

Software Product-Oriented Models

Software product models can be divided based on several perspectives. In one classification approach, software products can be classified based on the nature of the software into the system, support, and application software. In some other naming convention, system and support software are called upstream software while application software is termed as downstream software. Based on the platform, software products can also be classified into Web, OS, and mobile applications. Several software vendors are trying to offer the same product in those three different platforms. While major functionalities can be the same, however, each application environment has its constraints on the user interface, input, and output interactions.

The programming language used to realize software systems is also another variable used for classification. Software products are either written by high-level programming languages (e.g., Java, C#, C++, etc.). Scripting languages (e.g., Javascript, VB script, PHP, ASP, etc.) and in some cases assembly or machine level could be used to realize software applications. From software project management perspective software products can be classified based on size or complexity. For example, Constructive Cost Estimation model, COCOMO divided software products into organic, semidetached, and embedded based on size or complexity. Based on the based model, software products can be classified into commercial, open-source, and free software. Software products can also be classified into generic (for the general market) and custom (for a specific company or sector) products. Finally based on the product domain software applications can be classified into several types such as multimedia and graphics, business, enterprise, communication, educational, games, utilities, science and engineering, database, APIs, and several other types. A software development model that works well for revenue-oriented commercial software should be different from a software project built-in the open-source community. In Section 4 of this chapter, we elaborate on the unique elements for two types of software products: open-source and open competition projects. We analyze those types based on the factors that affect the software development process, and that should be considered when managing the development of such products.

Software People-Oriented Models

A software people model is a software development methodology that is centered on the people and their roles. In people models, we list individuals or roles, e.g., Analyst, GUI designer, architect, software developer, tester, etc. and elaborate on activities that each one of those should perform. In the following cases, we highlight the scenarios to demonstrate when a development model should be better people-oriented rather than process-oriented.

Software Project-Oriented Models

Similar to people-oriented models, in software project development models, the project type and nature should be important factors to affect project management decision making. Similar to software products, software projects can be classified based on different perspectives. In one view, they can be classified according to product classification. They can also be classified based on size and complexity. Software projects can also be classified into new/reengineering/maintenance projects. Each one of those projects has different expectations of success, revenue, resources, testing, etc. Each one also requires a different type of roles or skills. Generic software development models may not apply to all those types of projects as such generic models are mainly meant for new software products. Based on the amount of work from externals, software projects can be classified to in-house, partial outsourcing, or completely buy/hire a company to develop the product. The company may have the right to own and modify the code permanently or may have only the right to use the software (temporary or permanently). Based on risk and criticality, software projects can also be classified into critical and uncritical systems. Such criticality may be defined by the company or the user and based on different aspects such as money, health, level of accuracy or testing, importance, etc. Similar to products, software projects can also be classified in generic software projects to develop custom products.

Evolution of Software Process-Oriented Models

Initially, computers were mainly used for solving mathematical and scientific problems where algorithms were well-defined, and human involvement was minimal in selecting the workflow of the programs. Programmers employed computers to solve their problems, writing their instructions in machine-readable programs. Following assembly language was used as a programming environment followed by structured and object-oriented paradigms. Furthermore, the evolution of mini and personal computers meant that software systems were required in other domains, especially for office automation and educational purposes. This resulted in complex software internals, thereby increasing the complexity of its development activity too. As a result, now software applications were not only transforming mathematical algorithms into computer instructions but also supporting social practices in efficiently carrying their work. There was no standard development methodology employed to carry out software development, and as a result, most of the software projects turned failure, leading to the software crisis. As a result, the concept of software process emerged. Software process could be defined as the methodology to realize a software product. Humphrey [3] described the software process as a set of tools, methods, and practices used to produce a software product. Pressman [4] defined software process as a three-phase activity, i.e., definition, development, and maintenance. The definition phase focuses on extending requirements, whereas the development phase transforms those requirements into the software system, and finally, the maintenance phase aims at corrective actions and enhancements in the software system after release.

The oldest software process model is the waterfall model, which advocates for carrying out developmental activities in the following way (cf. Royce [5], Pressman [4]). This model works well when requirements are well known and fixed, which is very rare in real-world projects. In order to improve its limitations prototyping (cf. Bischofberger and Bomberger [6]), spiral (cf. Boehm [7]), incremental model (cf. Larman and Basili [8]), V model (cf. Pressman [4]), Rapid application development (cf. McConnell, [9]) and object-oriented paradigm (cf. Booch et al. [10]). Rational Unified Process (RUP) is an object-oriented model proposed and used by IBM, which stresses using the Unified Modelling Language (UML) design elements and tools developed by IBM. The model divides the process activities into micro and macro levels based on the division of the whole project into several modules (cf. Kruchten [11]). To reduce the development time, and increase flexibility, the notion of agile development emerged. These approaches accommodate changing requirements throughout the development process. The claim is that software projects are evolutionary by nature and the lifetime or the feasibility of the requirements for a software project may themselves be changed through the software development process. Agile development itself is not a software development process, but it is an approach advocating short iterative cycles, active user involvement and limited documentation (cf. Boehm and Turner [12], Schuh [13]). Some of the software process methodologies which adhere to agile principles are following.

- Scrum (cf. Schwaber and Beedle [14]).
- Feature-driven development (cf. Palmer and Felsing [15]).
- Crystal methodologies.
- Adaptive software development.
- Dynamic system development methodology.
- Extreme programming (cf. Paulk [16]).
- Lean software development.

Open-source and global software development approaches, where work is carried out by different distributed teams have posed new challenges for software engineers, as traditional software development approaches are not optimal in this scenario. The major highlight of these development methodologies is computer-based communication in contrast to face to face meetings [17, 18].

There are several examples of development models proposed for free or open-source projects. Cathedral and the Bazzar is one of the models which is usually mentioned as a typical example in open-source development models or methodologies. The model appeared for the first time in 1996–1997 by Eric Raymond. The two words represent a combination of two models: Cathedral from the commercial world, and Bazaar from the Linux world. Rather than having specific activities and tasks, this model focused on some guidelines on open-source development process [19].

Proposed a framework for open-source development. They suggested that such a framework should be able to systematically guide the creation and management of such communities within an organization. Such an approach can be accomplished through three steps: community building, governance, and infrastructure. The main idea is to have a light framework for communication to facilitate the flow of information among the source code development community. A similar idea is discussed by Madey et al. [20] to enforce utilizing social networks in the open-source development projects. It will be essential to extend source code version controls such as subversion, source safe, etc. to an Internet version (i.e., Distributed revision control systems). Such an extension is expected to occur in the open-source community shortly. Despite these research efforts, there is a need for a process model that could support open-source and outsourced software development processes.

K0131: Knowledge of Webmail Collection, Searching/Analyzing Techniques, Tools, and Cookies
K0140: Knowledge of Secure Coding Techniques
K0178: Knowledge of Secure Software Deployment Methodologies, Tools, and Practices.

Common Software Security Design Flaws

Security tools that check for vulnerabilities usually found vulnerabilities related to code implementation. Design flaws can be considered as roots or indirect cause too many of code vulnerabilities. For example, lack or improper methods of input validations are design flaws that cause many security vulnerabilities related to buffer overflow, SQL injection, etc. Similarly, access control or encryption design flaw or improper design can create different types of vulnerabilities and exploits. Literature includes several reports on top or most common or essential software security-related design flaws to avoid. The following is a summary list:

- **Validate user inputs**: All software applications require users' interactions to respond and include their inputs. User inputs can come in different forms from the very free text form (where users can type anything they won't) to some restricted forms where users will pick from. Ideally, and whenever possible, limit the amount of data to input by users.
  If possible, make all or most of those as select from alternatives where users will not have to type any free text. However, if functionalities require users' free input text, validate this input and don't assume users are going to follow generally accepted roles. Users intentionally or unintentionally may try to violate those roles. Validate user input as early as possible. There are many possible designs.
  In some cases, especially in web applications, users can be prevented, at the user interface level, from trying to include invalid inputs or characters. For example, if the form does not accept more than five characters, even if users try to type more

than five characters, the user interface will not allow them. As an alternative, inputs can be validated in the code rather than the user interface. Mainly, this input should not be further processed by code and certainly not be allowed to reach back-end databases, without being properly validated.

In some types of Man in the Middle attacks (MiM), attackers may try to attack and tamper data in transit. This means that proper input data validation should not be only employed to inputs directly taken from users, but to data inputs in general. Some web applications may receive inputs from other websites (e.g., a bank, credit agency, government agency, etc. website). Input data should be validated from both trusted and untrusted sources.

- **Use encryption properly or correctly**: Encryption should be used whenever possible. Encryption is used in many different places in software applications and information systems. It is used to hide information in rest (i.e., in files and databases) and is also used to hide information in transit. In some scenarios, encryption is required to hide all communication content between two parties. In some other scenarios, only some parts of this communication (e.g., users' credentials) should be encrypted when in transit. Use the correct implementation of encryption as merely using encryption does not guarantee data, identity, and privacy protection.

- **Use proper authentication and authorization levels**: Software developers usually use and employ security control modules and APIs. Those are integrated within the environment and can be used across different applications. Software developers do not need to reinvent the wheels and in most cases, properly implement whatever security standards exist. Security control mechanisms exist in operating systems, access control systems, network components, database management systems, websites, software applications, etc. As such, the application security is one layer in this layered architecture that should be designed in alignment with this overall architecture.

  In some cases, a software application may not need to consider security modules internally but properly integrate with security components that exist in the environment. In security layered architecture models, authentication and identity management comes before access control or authorization. Authentication tries to answer a yes, no question whether user, application, request, etc. can be allowed or blocked access request or entry. Once allowed access control mechanism should make a further decision of what permission levels this request can have.

  About this part, code for data processing should be separated from that of security control processing. Data can be manipulated by users or their input, and hence this data should not be mixed with control to commit attacks such as those related to privilege escalation ultimately.

- **Design for future changes**: Many of the design problems arise when a program design is in continuous change. Initial designers may do a good job based on the initial scope. However, there was no clear plan for extensions. Extensions/modifications for software applications are inevitable. They will come either as part of maintenance, bug fixes, etc. They may also come in response to new

```
while(){
  ...
  try{
   case...
   default:
     throw new IllegalArgumentException("Invalid input...");
  }catch(IllegalArgumentException iae){
   //do stuff like print stack trace or exit
   System.exit(0);
  }
}
```

Fig. 7.3 An example of exception handling

features or due to changes in the environment where the software must accommodate those changes. Cosmetic design changes that may come in response to
those changes may ignore essential design principles.
- **Take careful and extra considerations when dealing with sensitive data**:
  Attackers start from weak or vulnerable targets. Software designers should do
  their best to harden their applications. Sensitive data can be easily located in
  any application by software developers as well as attackers. Hence, designers
  should put extra considerations for areas that include sensitive data and make
  sure through, frequent auditing and testing that those sensitive areas are not easy
  targets to attack.

Software secure construction and defensive programming, exceptions and error handling. Software code exception handlings are code components that are called in
unusual, exceptional cases. The main goal of design for code exceptions is to stop
software applications from crashing at run time. Exception handling techniques can
be used to block invalid inputs from users. Figure 7.3 shows a simple example of
code exception handling.

Exceptions block invalid user inputs early in the code before further code processing and before passing those inputs to internal system components (e.g., Databases).
Exceptions also serve an important goal of communicating effectively with users to
show them what went wrong or how to communicate with the application with valid
inputs properly. This is an important "usability" software quality attribute. Users
should know why their requests are not accepted or processed. Exception feedback
should help users fix the problems with their inputs. However, exception feedback
should not be "revealing." In many security attacks, attackers intentionally try different types of inputs and try to use feedback exceptions to learn more about the
system. Recent security research projects focused on this issue and why software
exception handling feedbacks need to balance between usability and security issues.
Balance feedback ensures that exception handling can serve both software quality
and usability goals.

Software Malware Analysis

Malwares are software applications built with malicious intents. They are sent to user desktops, laptops, mobiles, or websites through different techniques such as tricking users into receiving or downloading those Malwares through some social engineering techniques. Anti-Malware software applications can work either per request or in real-time to detect such Malwares, alert users, and block them from accessing computing resources.

Malware analysis can be conducted automatically by anti-Malware systems or manually by security and forensic experts. Due to performance and speed issues, anti-Malware systems employ lightweight, quick detection methods in comparison with those sophisticated detection and analysis techniques employed through manual analysis.

Anti-malware Detection Techniques

The following three primary detection techniques are employed by anti-Malware systems to detect Malwares:

- **Signature-based**: Anti-Malwares used simple and reliable signature-based methods to identify suitable applications and files from malicious ones. A popular signature-based method is hashing. Hashing algorithms such as SHA1 and MD5 can be used to generate unique Hexadecimal values for system files and folders. Figure 9.2 shows a simple file hashing example with values from different hashing algorithms.
- **Role-based**: Hashing-based Malware detection can work well in known territories. However, for new Malwares, such Malwares will not be registered as known malicious files or applications. Role-based detection methods are based on the identifications of different behaviors that indicate that the subject file or application is Malware. For example, an application that tries to access or change certain sensitive areas or information in operating system kernels can be suspected as a Malware. This may cause a significant number of false positives, where standard applications are trying to trigger such requests. It may also cause cases of false negatives where some Malwares may not be detected as they are either trying to make hidden moves or their triggers are not recorded by any defined role.
- **Behavior analysis**: Malwares can be very complicated. Detecting such complex Malwares may take more than a second or a part of a second in real-time. In behavior analysis, a specific sequence of activities may trigger a suspicious behavior.

In some cases, this may require full packet states' inspection (Also called deep packets inspection) in order to be able to judge whether the subject is a Malware or not. Particular library (e.g., APIs) or system calls can also be defined as Malware-like behaviors. Anomaly detection methods define ranges of acceptable behaviors. Any "deviation" from such behavior can be classified as "anomaly" and hence by a possible Malware.

K0212: Knowledge of Cybersecurity-Enabled Software Products

The term cybersecurity-enabled software product indicates that the software product complies with all recent security regulations and standards. Different security and quality checks take the process in the different stages of the software development lifecycle to make sure that software is not vulnerable by itself and will not cause the underlying network, operating system, etc. to be vulnerable. Due to the comprehensiveness of the subject, we will focus on samples of actions taken toward this goal.

Encryption in Operating Systems and Disks

Disk encryption is used to encrypt every bit of data that is written to the disk or disk volume. This is used based on disk encryption software or hardware. Disk encryption is used to prevent unauthorized access to data storage. Disk encryption provides another layer of information protection (beyond the network, operating system, etc. access control). Disk encryption is also different from file encryption that provides encryption per files as disk encryption software provides encryption for the whole disk, bit by bit, as one unit.

For file encryption, Microsoft used the Encrypting File System (EFS) in its NTFS file system (Version 3.0). Files can be encrypted and protected from users who can have access to the desk or the operating system. EFS uses symmetric encryption algorithms which can be faster to encrypt and decrypt in comparison with asymmetric algorithms.

Followings are examples of popular disk encryption software:

- **BitLocker for Windows**

Microsoft started using BitLocker encryption in its operating systems from Windows Vista and later on. The system is using AES encryption with CBC or XTS (cipher-text stealing) mode.

There are two types of attacks to consider here: The attacker may try to know the information in the disk, and we want to prevent them from doing so. Second, the attacker may try to tamper such information, without being able to know it, and we also want to prevent them from doing so.

Using Elephant diffuser is Microsoft's proposed solution to CBC mode modification to allow it to prevent data tampering, the specific data tampering and not the random. The idea here is to "mix" the plaintext up before encrypting it so that an attacker can't change specific data (called data meddling or malleability). Microsoft then removed Elephant diffuser from being the default cipher choice in Windows 8 and beyond (citing speed and standard compliance as reasons). The mode is kept for backward compatibility.

XTS is introduced in Microsoft operating systems in Windows 10. In comparison with CBC, there is no requirement for an initialization vector, IV for XTS. XTS tweak key can be derived from the block number. This tweak key can be sector address or a combination of the sector address and its index.

Each block is encrypted differently based on the different tweak values. Additionally, in XTS each AES input is XORed with a differently shifted version of the encrypted tweak. This can prevent an attacker from changing one specific bit in the disk.

It is argued then that while both CBC (undiffused) and XTS can prevent the first attacking type, XTS only is capable of preventing the second attacking type. This is one of the reasons for considering diffused CBC mode. XTS-AES is used by full-disk encryption systems, such as LUKS, TrueCrypt, FileVault, and Microsoft BitLocker.

XTS incorporates two AES keys, K1 and K2. The first key: K1 is used for encrypting the sector number to compute per-block tweak values. The second key: K2 encrypts the actual data.

While XTS has its limitations and some performance issues, currently, it is considered as an excellent choice.

- **FileVault for Apple OS/X**

Apple Mac's FileVault uses XTS-AES-128 encryption that we described earlier. FileVault, used with Mac OS X Panther 10.3 and later, allows you to upload a copy of your recovery key to Apple so you can recover your files via your Apple ID if you ever lose your password. This is optional if the user decides to store the key in iCloud.

FileVault 2 was introduced in Mac OS X 10.7 ("Lion"). Unlike FileVault 1, that encrypts only user data, FileVault 2 encrypts the whole disk Full-disk encryption (FDE).

One disadvantage of using FileVault 2 is that it uses the user's Mac OS password. It is generally recommended to use multi-factor authentication, rather than same OS credentials.

- **TrueCrypt**

This is a non-commercial freeware encryption application, for Windows, OS/X, and Linux (www.truecrypt.org, www.truecrypt.sourceforge.net). The project started in 2004. TrueCrypt used LRW and XTS modes of encryptions. The project was terminated in 2014 and is no longer considered safe to use for disk encryption. The project faced some legal problems and disputes since it started. The disk encryption was abandoned from Windows operating systems beyond Windows XP, and similarly from recent Linux and Mac operating systems. Two alternative projects came to replace TrueCrypt: VeraCrypt and CipherShed.

VeraCrypt uses 30 times more iterations when encrypting containers and partitions in comparison with TrueCrypt. This means it takes more time to perform the encryption/decryption process.

VeraCrypt (https://veracrypt.codeplex.com/) has released version 1.21 in July 2017 with many security improvements in comparison with TrueCrypt. VeraCrypt uses XTS mode and supports AES encryption as well as some other algorithms: Serpent, Twofish, Camellia, and Kuznyechik.

To balance between security and performance, VeraCrypt supports parallelization in the encryption/decryption process.

- **ESSIV**

Encrypted salt-sector initialization vector (ESSIV) method for full-disk encryption is proposed to "replace" CBC mode. In particular, ESSIV proposed a new method to generate.

Initialization Vectors (IVs). This is as the usual methods for generating IVs are predictable sequences of numbers based on a timestamp or sector number which may permit certain attacks. IVs are generated from a combination of the sector number (SN) with the key hash. SN combination with the key in the form of a hash makes the IV unpredictable. ESSIV is used by different versions of Linux since 2000.

To promote secure practices in IT and software products and devices, the NSA recommends the following:

1. Upgrade to a Modern Operating System and Keep it Up-To-Date
2. Exercise Secure User Habits
3. Leverage Security Software
4. Safeguard against Eavesdropping
5. Protect Passwords
6. Limit Use of the Administrator Account
7. Employ Firewall Capabilities
8. Implement WPA2 on the Wireless Network
9. Limit Administration to the Internal Network.

Security-Enabled Web Browsers

A security-enabled browser can confirm the identity of the Web site before transmitting and even notify users if it looks suspicious. This helps users in predicting and eliminating many of the candidate phishing attacks. Websites that claim they are secure or encrypted, and they are not can be detected by security-enabled browsers and alert users of such situations.

Users should be aware of the security features available through their browsers. They should use a security-enabled browser to specially to make purchases and access their information-sensitive websites. For example, if you do not use an SSL-capable browser, you are at risk of having data intercepted. Transport Layer Security (TLS) and is the protocol which almost all security-enabled browsers to use. It was designed by Netscape and was formerly known SSL or Secure Sockets Layer.

K0236: Knowledge of How to Utilize Hadoop, Java, Python, SQL, Hive, and PIG to Explore Data

Data analytics projects have a lifecycle that contains the following major stages:

- Initial analysis and planning: and direction: Similar to any project first stage should include requirements analysis and planning. We should have defined goals or else data analysis, and intelligence will be very time-consuming and unfocused. The process can be, however, evolutionary where initial requirements and plans can be a good start (in the first cycle). Outputs from earlier cycles can be used to improve further analysis and planning in the next cycles.
- Data collection stage: Data is collected, manually, or through tools from the different sources, we have mentioned earlier. Programming and scripting languages such as Python, R, Ruby, Java, Go, etc. can be used to automate the parsing process. Many websites may resist the parsing or crawling process (especially OSNs). Alternatively, those websites offer their APIs (mostly with limited capabilities) to parse their data (e.g., see: https://developers.facebook.com, https://dev.twitter.com/docs, https://www.npmjs.com/package/google-trends-api, etc.). Data can also be collected from logs such as Honeypots, Firewall logs, Intrusion Detection System logs, scans of the Internet, etc.
- Data processing—Several data preprocessing techniques are typically employed in the data analysis activities. For example, this stage may include how to prepare data for analysis (e.g., stemming, stop-words' removal, etc.), data storage, and retrieval methods. In some cases, data can be stored into a text file, small scale databases, or big data repositories.
- Data analysis and production: This is the main goal and most time-consuming task in the cycle. In this task, knowledge, and intelligence, according to the project goal, are extracted.
- Data dissemination and usage: In an evolutionary process, this can trigger further data analysis in future cycles. In later cycles, knowledge and intelligence are produced to decision-makers or target audience.

K0279: Knowledge of Database Access Application Programming Interfaces (APIs) (e.g., Java Database Connectivity [JDBC])

Application programming interfaces (APIs) are small or micro-programs that are typically designed to provide generic services to a spectrum of users or applications. They can also be used to work as interfaces between users and their applications, between applications and operating systems, between applications and the data or the database management system (DBMS) or between all those and external systems, hardware devices, etc.

Database APIs like JDBC are used to enable the connection between user-created applications and DBMS to connect to and eventually the database. For example, in windows, users can download the right JAR for JDBC, (e.g., MSSQL-JDBC-7.2.2.jre8.jar, and MSSQL-JDBC-7.2.2.jre11.jar class library files) for MS SQL server. They can then use the right database source within the installation of the JDBC instances (Fig. 7.4).

K0328: Knowledge of Mathematics, Including Logarithms, Trigonometry, Linear Algebra, Calculus, Statistics, and Operational Analysis

Discrete math may provide creative programming challenges. Similar to engineering, programming is an art and a science. A programmer needs to have a strong mathematical background. However, without the talent to write the proper code, this strong mathematical background will not be enough. A programmer may develop an algorithm heuristically to solve a complex linear or integer complete problem despite the fact that the mathematical background will help programmers write efficient and smart algorithms. A programmer who wants to solve a simple linear equation such as $A = 5B + 18$, will define two variables, A and B. A user will then select a value for B and ask the program for the result, or B. Such simple problems can be transferred directly from math to programming. However, this is not the case for the majority of problems we will try to solve in this book.

Basic Algebra

To solve complicated mathematics problems, we need analytical thinking, an algorithm or algorithms of possible ways for solving problems, and some experience with the field. Mathematicians may find different ways to solve the same problem. Better mathematicians are those who solve problems efficiently and, in a format, that others can easily understand. In the same standards, programming involves the same set of

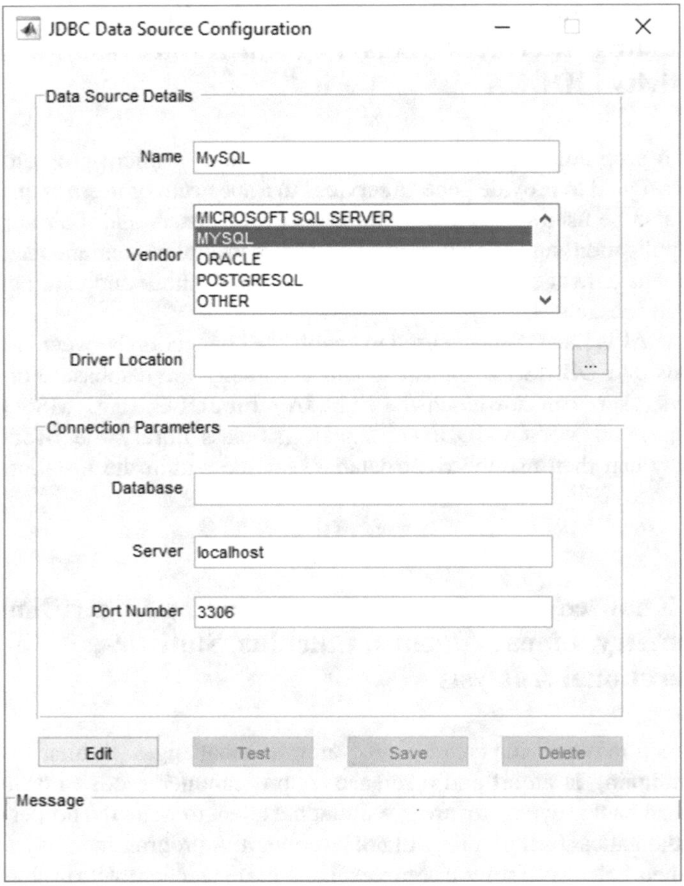

Fig. 7.4 JDBC connection with the data source

skills. In programming, the same set of principles applies to the solving of problems. When you have the right standard solution to a programming dilemma, this solution is included in a library that can be reused in other or more significant problems later. These skills seem to be very similar to the skills used in mathematics. Programming structs and languages have much more in common than those of math.

In programming languages, mathematical expressions can be built and evaluated. The Table 7.2 shows the operators that are supported by programming languages in general to perform mathematical equations.

Let's first try to solve the expression below, Fig. 7.5, using language notation. First, we need to know the precedence concept between operators. In any programming language, operators are prioritized, once they come together with no parenthesis. Below you'll find a table illustrating this prioritization. Here in the two lines, equations are solved directly from left to right as operators have the same precedence (Table 7.3).

Table 7.2 Operators representations in programming languages

Operator	Description	Algebra	Language
+	Addition	$x + y$	$x + y$
−	Subtraction	$x - y$	$x - y$
/	Division	x/y or $x \div y$	x/y
*	Multiplication	$3x$ or $3 \cdot x$	$3 * x$
%	Modulus	$x \bmod y$	$x \% y$

Fig. 7.5 Simple mathematical expressions

$$3 + 2 - 5 \cdot 30 - \frac{5}{15};$$

Table 7.3 Operators precedence in programming languages

Group	Associatively	Example
*, %, /	Left to right	$5 \div 10 * 12 \div 8 := 0.75$
+, −	Left to right	$3 - 4 + 5 - 6 := -2$

This equation will be solved by first performing the multiplication and division operations from left to right concerning their positions in the equation. Next, subtraction and addition operators are executed in the same way; 5 * 30 gets evaluated first, 5/25 next. After that $3 + 2$ is evaluated and the result of $3 + 2$ is then subtracted from the result of 5 * 30. The result of the computation is then subtracted from 5/15. The final result is -145.332.

This may look cumbersome (if it is computed manually). A computer program will not have a problem understanding such a technique. However, to avoid such ambiguity, you can enforce the precedence you want through using parenthesis to apply direct computation of what's within the first. For instance, in the equation:

$$(3 + 2) - 5 \cdot 30 - \frac{5}{15};$$

$3 + 2$ is evaluated first. Having parenthesis does not violate the precedence rules. It simply overrides it. Evaluate always what is inside the parenthesis first and then follow whatever precedence rules are telling you to do. In the case of nested parenthesis such as $(x + (y/z))$, the innermost pair of parentheses is evaluated first, and so on.

Table 7.4 summarizes the precedence roles of mathematical operators.

Table 7.4 Mathematical operators' precedence in programming languages

Operator(s)	Description
()	Parenthesis
*/%	Multiplication, division, modulus
− +	Addition, subtraction

Table 7.5 Some examples of built-in mathematical functions in programming languages

Function	Description	Mathematical
$\cos(x)$	Cosine of x	$x := 5.0 \rightarrow \cos x := 0.283662;$
$\sin(x)$	Sine of x	$x := 5.0 \rightarrow \sin x := -0.958924$
$\tan(x)$	Tan of x	$x := 5.0 \rightarrow \tan x := -2.38052$
$\text{sqrt}(x)$	Square root of x	$x := 25 \rightarrow \sqrt{x} := 5$
$\text{pow}(x,\ n)$	x raised to the power of n	$x := 5, n := 2 \rightarrow x^n \rightarrow 5^2 := 25$
$\text{ceil}(x)$	Cast x to largest integer	$x := 5.2 \rightarrow \lceil x \rceil := 6$
$\text{floor}(x)$	Cast x to smallest integer	$x := 5.9 \rightarrow \lfloor x \rfloor := 5$
$\text{abs}(x)$	x absolute value	$x := -5 \rightarrow |x| := 5$
$\text{fabs}(x)$	Floating x absolute value	$x := -5.6 \rightarrow |x| := 5.6$
$\log(x)$	Natural logarithm of x	$x := 10 \rightarrow \log x := 1.60944$

Sometimes you may need to perform some functions on numbers. Programming languages provide a large library of built-in functions. Below is a concise list of such functions (Table 7.5).

Programming built-in mathematical functions are somehow limited but yet powerful. You may sometimes need to construct your functions to solve a given scenario. For instance, $\sum_{i \rightarrow 1}^{n} |x_i|$, $\prod_{i \rightarrow 1}^{n} |x_i|$, P_k^n and C_k^n. We will be using such functions intensively for solving many mathematical equations.

In algebra, assignment statements are straightforward. Programming languages provide shortcuts for such assignment statements. Table 7.6 shows an example of such shortcuts.

Table 7.6 Assignment shortcuts in programming languages

Algebraic	Language
$x = x + 2$	$x\ += 2$
$x = x \cdot 3$	$x\ *= 3$
$x = x/5$	$x\ /= 5$
$x = x - 10$	$x\ -= 10$
$x = x \bmod 2$	$x\ \%= 2$

Operator	Description	Algebraic
Table 7.7 Equality and relational operators		
==, !=	Equality	$=, \neq$
>, <, >=, <=	Relational	$>, <, \geq, \leq$

Equality and Relational Operators

Sometimes you need to have an operator that yields a yes/no result. This type of operator will be helpful in conditional or logical programming practices. Table 7.7 shows the equality and relational operators.

Let's illustrate their usage in the following scenarios:

Let's *have three variables x, y, and z. X should be assigned to y if z is greater than 10 and y equals to 5, else x should be assigned −10.*

We *have an array of integers, and we need to know how many of them are greater than or equal to 10.*

An array of integers should only contain numbers ranging from 10 to 20.

Equality and relational operators are used to enforce constraints on processing the data.

Principles of Logic

In computers, logic can be derived from algorithms intended to ensure information integrity and reliability. To fully understand logic, we need to observe human behavior in making decisions or conclusions. Humans have the ability to draw conclusions and formulate inferences based on specific patterns. Patterns are those well-defined knowledge objects that are somehow derived from trial and error or by learning techniques.

For instance, humans should be able to conclude that the probability of getting injured as a result of falling from a high building is high. The question is, then, how did they know? Human knowledge is a result of accumulated information we gathered through life, experience, others, etc.

As mentioned earlier, humans learn and store knowledge about specific behaviors and phenomena. The ability to retrieve such relevant information is actually considered a logical operation as it stands on the top narrow level of the knowledge pyramid. Logical thinking means that the output of such operations is a decision (i.e., yes/no or true/false, etc.). Logical statements may include many logical operators, for the statements to have a meaning.

Logical operators are used to joining propositional statements and allow greater flexibility in evaluating multiple statements (Table 7.8).

For an AND statement to be true, all operands need to be true, for example. To be qualified for graduation, a student Grade Point Average (GPA) should not be less than 2.0, and he/she should complete at least 134 h of study. He/she will not graduate

Table 7.8 Logical operators in programming languages

Operator	Language	Mathematical
AND	&&	$\wedge$
OR	||	$\vee$
NOT	!	$\sim$

unless both conditions are fulfilled. On the other hand, the OR operator requires only one of its operands to be true for the result to be true (For example, to register for class CS360 you should have taken CS310 or CS320. One of these two requirements should be enough to register for CS360. It will not be a problem if both are achieved (Table 7.9).

The NOT operator negates a statement, for example; a student can register for a class if that class has not reached its maximum limit. The statement can be restated as "if the class is NOT full, then the student can be enrolled in it.

Logical operators are joined to allow for more complex propositional statements. For example, a student who wants to register for class CS330b should have taken CS320 and CS203 or CS204 or CS205 and not CS330a. In this statement, precedence rules should be taken into consideration, the NOT operator has the highest priority, the AND operator comes next, and finally, the OR operator gets evaluated.

From a design, perspective try having relevant operators grouped by parenthesis to avoid logical errors, and for clarity, logical errors occur when the software computes something other than expected, for instance in the above example the software will first check whether the student has taken both CS320 and CS203 but, in fact, it should evaluate whether the student has taken CS320 and at least one of (CS203, CS204 or CS205) and not CS330a (Fig. 7.6). In this case, the propositions included in the parentheses are evaluated first from left to right because as stated earlier, parentheses

Table 7.9 Truth table for logical operators

A	B	$A \wedge B$	$A \vee B$	$\sim A$
True	True	True	True	False
False	True	False	True	True
True	False	False	True	False
False	False	False	False	True

If Exists ("CS320") and (Exists ("CS203" or "CS204" or "CS205")) and not Exists ("CS330a") then

    Message "Student can register"; Else:

    Message "Student cannot register"; End if

Fig. 7.6 Logical statements with several operators

If Exists ("CS320") and not Exists ("CS330a") then ...

*If Exists ("CS320") and true and **true** then ...*

*If (**true** and true) and true then ...*

.   *If (**true** and true) then ...*

*If (**true**) then ...*

Fig. 7.7 Function execution

have the highest priority. Next, the AND statements are evaluated from left to right, and so on.

In the statement: "A student took the courses: CS320 and CS204 but not CS205 and CS330a, the student is registering for class CS330b", should this student be allowed to perform the registration? Why?

When the function evaluates the propositions (trace the function execution in Fig. 7.7), it first checks whether the student took any of (CS203, CS204, or CS205 courses) because they are contained within parentheses. The statement then holds, and it's replaced by true (a). Next, it checks the NOT statement and replaces it with true since the student did not register for class CS330a (b). Next, the function evaluates the last proposition, where it checks whether the student took the CS320 course and replaces it with true since the student did register for that class (c).

As stated earlier, the AND operator requires that both of its operands hold in order to pass. As a result, the leftmost AND operator returns true (d). Next, the last AND operator also returns true (e). As a result, the proposition is satisfied, and the registration continues. Try to solve the problem without grouping the operators.

Logical, relational, and equality operators are meaningless without logic gates where the resulting inferences are evaluated to draw conclusions. In the earlier example, registration was granted as a result of satisfying the conditions (propositions), where propositions are a result of satisfying logical, equality, or relational conditions. In programming, logic gates can be constructed by using if and switch statements. If and switch statements are used for simple as well as complex problems. The switch statement is used only with basic propositions, whereas if statements are used with different kinds of propositions. However, any switch statement can be restated as a group of several if statements (Fig. 7.8).

Several if statements can be implemented inside each other. This is then called "nested if statements" where an (if) statements will include one or more if statements inside. Each one of the inner "IF" statements may also include one or more statements. For instance, the above example could be restated as follows (Fig. 7.9).

Nested if statements help execute more detailed explanations and clarifications. When propositional statements get large and sophisticated, the statements should be divided into multiple statements considering their independence and priority. For instance, in the earlier example, the CS320 has the highest priority because it is a direct prerequisite for the CS330b class.

If x_i < 10 Then: *Message "x is less than*	*Switch (x_i)*
10";	*Case is less than 10: Message "x is less than*
Else if x_i ≥ 15 and x_i ≤ 20 Then	*10";*
Message "x is between 15 and 20";	*Case is between 15 and 20:*
Else: Message "x is larger than 20";	*Message "x is between 15 and 20";*
End if	*Default: Message "x is larger than 20";*
	End Switch

Fig. 7.8 Similarities between the IF and SWITCH statements

If Exists ("CS320") then
    If Exists ("CS203" or "CS204" or "CS205") then
        If Not Exists ("CS330a") then
            Message "Student can register";   *Else*
            Message "Student cannot register, CS330a is taken"; End if
        Else: *Message "Student cannot register CS203 or CS204, or CS205 should be*
taken"; End if
        Else: *Message "Student cannot register, CS320 should be taken"; End if*

Fig. 7.9 Nested if statements

Linear Programming

Linear programming is an important tool for decision making. Once we model a problem into mathematical equations, the goal of linear programming is to find the best solutions for the problem that fulfill its constraints. Constraints are considering pass/fail decisions where those constraints should all be fulfilled for a solution to be feasible. Unlike the variables and the equation, itself, constraints will have only yes or no decisions. Either the constraints are fulfilled, which will make us consider the solution, or drop the proposed solution. Those are similar to the pre-requisites or preconditions that are required before getting into the execution of a solution.

A linear equation is an equation in which the highest degree for a variable is linear. For example,

$$Y = X + 1,$$
$$A = 3b + 2c + 3d + 5,$$
$$G = 6d - 3A + 2C + 5X.$$

In linear programming, both the objective function to be optimized and all the constraints, are linear in terms of the decision variables. An objective function is the main function that we are trying to optimize. The constraint equations are all limitations or regulations that control or limit the boundaries of the solution.

Linear programming in its standard form consists of three parts:
It consists of three parts, which are given as follows:

1. A linear function to be maximized:
   $$\text{maximize } c_1x_1 + c_2x_2 + \cdots + c_nx_n$$
2. Problem constraints
   subject to $a_{11}x_1 + a_{12}x_2 + \cdots + a_{1n}x_{1n} \leq b_1$, b_1, $a_{21}x_1 + a_{22}x_2 + \cdots + a_{2n}x_n \leq b_2 \cdots$

$$a_{m1}x_1 + a_{m2}x_2 + \cdots + a_{mn}x_n \leq b_m$$

3. Non-negative variables

$$\text{e.g., } x_1, x_2 \geq 0$$

The matrix form:
The problems can be usually expressed in matrix form and then it becomes:
   maximize $c^T x$
   subject to $Ax \leq b, x \geq 0$

LP Formulation

$$\text{Maximize } 5x_1 + 7x_2$$
$$\text{Suchthat } x_1 \leq 6$$
$$2x_1 + 3x_2 \leq 19$$
$$x_1 + x_2 \geq 8$$
$$x_1, x_2 \geq 0$$

Standard Form

$$\text{Maximize } 5x_1 + 7x_2 + 0s_1 + 0s_2 + 0s_3$$
$$\text{Suchthat } x_1 + s_1 = 6$$
$$2x_1 + 3x_2 + S_2 = 19$$
$$x_1 + x_2 + s_3 = 8$$
$$x_1, x_2, s_1, s_2, s_3 \geq 0$$

The feasible region (Fig. 7.10), represents the region wherein the best solution should exist. This region is allocated through the constraint equations.

Fig. 7.10 Feasible regions

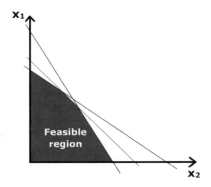

K0373: Knowledge of Basic Software Applications (e.g., Data Storage and Backup, Database Applications) and the Types of Vulnerabilities that Have Been Found in Those Applications

Websites such as NVD (https://nvd.nist.gov) and CVE (https://www.cvedetails.com/) continuously record security vulnerabilities based on a US national standard since the early 90s. In this paper, we are going to focus on discussing two issues with those vulnerability databases: The classification of those vulnerabilities under different categories called CWE (Common Weakness Enumeration) and the seriousness/criticality scores of those vulnerabilities CVSS. Examples of problems discussed in CWE classification are that they are (1) inconsistent, and (2) many categories are not exclusive which means that some vulnerabilities can be included in more than one CWE category.

The Common Vulnerability Scoring System(CVSS) provides a quantitative mechanism to reference information-security vulnerabilities. The CVSS' numerical score reflects the severity of exploits in a qualitative representation such as low, medium, high, and critical to helping prioritize the vulnerability management process. This Common Weakness Enumeration (CWE) classifications are inconsistent and inclusive in more than one of CWE's categories. The CVSS categorization focuses on the implementation issues such as XSS or SQL but overlooks the design and architecture of the software. For example, the impact of XSS flow receives "low" weight in the meantime; there is no input validation mechanism; in this case, the given weight can be elevated to medium or high. The priority of design flows should be considered in a higher category than implementation or use separate factors in categorization. The business impact and the use of products are different. A Safety and reliability product's scores should be evaluated based on business impact. The value of a score should be dynamic; the organizations should reevaluate the sore since the likelihood of a weakness frequently changes due to improved detection techniques and technology advancement. The frequency of these changes could impact the remediation process of a weakness whose category level has been reduced.

Trend Analysis of Vulnerability Types Since 2000

Figure 7.11 shows major vulnerability trends since 2000. The Figure shows the following major categories of vulnerabilities:

- Buffer errors;
- Permissions, privileges and access controls;
- Input validation;
- Cross-Site Scripting (XSS);
- Information leak/disclosure.

Figure 7.12 shows vulnerabilities in total. Some of the severe vulnerabilities such as buffer errors, access control problems, input validations, etc.

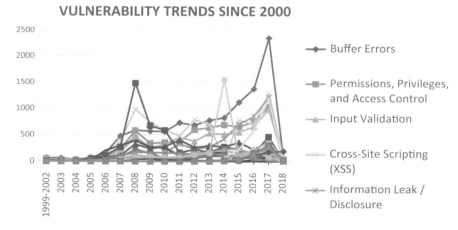

Fig. 7.11 Vulnerability trends since 2000

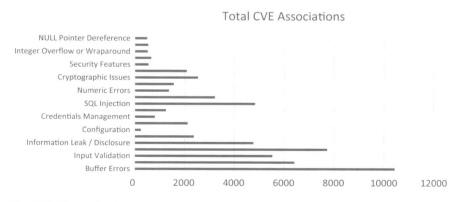

Fig. 7.12 Totals of major vulnerabilities since 2000

K0396: Knowledge of Computer Programming Concepts, Including Computer Languages, Programming, Testing, Debugging, and File Types

Core Object-Oriented (OO) concepts include:

- Objects/Classes;
- Data Abstraction;
- Modularity;
- Encapsulation;
- Inheritance;
- Polymorphism.

Key goals of OO software paradigm:

- Reuse: Why write the same or similar code again and again, eventually create more efficient code.
- Extension/Maintenance: Code should be designed to allow/enable/embrace future changes for whatever reason with minimum effort.

Classes Versus Objects

- Class are templates of objects; design once use many times.
- Objects are instances of classes that can have a state (fields), a behavior (methods), and an identity (names).
- Classes Versus Objects: Design Versus Implementation, The What Versus the How, Classes have the attribute types but not values.
- Objects as Entities can come from real-world objects, or not.
- Objects interact with each other.

Abstraction

- Design for what you need.
- Software design is all about abstraction.
- Abstraction converts all project needs/constraints, etc. into a unique solution prototype that can fulfill them.
- Suppress irrelevant implementation details.
- This depends on the project domain and context. For example, for a student class in a University system (weight may not be a relevant attribute to include, but for the same class in the University clinic, it is relevant).

Data and Method Abstraction (Interfaces)

- Include what is relevant and eliminate or hide what is irrelevant.

- A class interface is its public methods and attributes. Those should be generic and rarely change.
- On the other hand, everything else, concrete implementation, should be private, and hence can change frequently, will not impact class users.

Encapsulation and Information Hiding

- (Encapsulate what varies): Abstract Data Types (ADT).
- Interfaces (public methods and variables) should change independently from the concrete implementation (i.e., internal or private methods and variables).
- One thing that distinguishes OO languages is the heavy usage of ADTs or libraries.
- Levels of Abstractions/Details: Details in a project can be relevant/irrelevant in specific project modules, times, etc.
- Abstraction Leads to information hiding and encapsulation: In abstraction, you include what is relevant to your project, classes, etc. and eliminate what is irrelevant.
- Push to the public (i.e., interfaces whether they are class interfaces or user interfaces) only what you want others to see, use, and manipulate, encapsulation. Hide from users' irrelevant details; Information hiding.
- Information Hiding. Know only what you need to know. For example, a user who is looking for a sorting method, may not care or need to know the (sorting algorithm) as long as it sorts their input data correctly and generates sorted output data.
- Information Hiding. Know only what you need to know.

Modularity

- The aspect of syntactically grouping related declarations. (e.g., fields and methods of a data type).
- Package/classes and even methods, the location of all those in your project should come with modularity in mind (Remember OO two main goals increase reuse and decrease change).
- In OOPLs, a class serves as the basic unit for decomposition and modification. It can be separately compiled.
- Don't overdo/underdo grouping, the context of the project should decide (e.g., the number of classes in a project Vs. the number of methods in each class).

Coupling and Cohesion

Coupling and cohesion focus on designing software modules that can be interchangeable and that can be easy to reuse, modify, or eliminate in the future.

- (Decrease coupling and increase cohesion).
- Encapsulation: Controlling the visibility of names. Enables enforcing data abstraction.
- Conventions are no substitute for enforced constraints (Don't assume but enforce).
- Enables mechanical detection of typos that manifest as "illegal" accesses.

Inheritance

- Subclasses, Why? Three main goals: Reuse, extend, override.
- Code reuse: derive Colored-Window from Window.
- Specialization: Customization.
- Generalization: Factoring Commonality.
- Code-sharing to minimize duplication.

Classes Versus Abstract Classes

- Many OO design principles are related to the right use of inheritance.
- Open-closed principle: Extend but do not modify. A class is closed because it can be compiled, stored in a library, and made available for use by its clients.
- Stability: A class is open because it can be extended by adding new features (operations/fields), or by redefining inherited features.

Interfaces

- Primary mechanisms to achieve sound design principles.
- Most of new Java/C++ and C# libraries increase the usage of abstract classes and gradually moved them to interfaces.
- Why? What is the difference between abstract classes and interfaces?

Polymorphism (Many Forms)

- Polymorphism in OO promotes reuse in a dynamic environment.
- Popular in Gaming.
- Integrating objects that exhibit typical behavior and can share the "higher-level" code.
- Unifying heterogeneous data.

Binding: Static Versus Dynamic

- Static binding (resolved at compile-time).
- Dynamic binding (resolved at run time).

Class to Class Relations

- Four types: Inheritance, composition, aggregation, and usage/association.
- Composition (Client Relation) ("has a").
- Inheritance (Subclass Relation) ("is a").
- Inheritance is a stronger relation than composition/aggregation, better reuse, but more complexity.
- Composition Versus Aggregation: Composition requires/includes Physical Containment.

K0531: Knowledge of Security Implications of Software Configurations

Many software configuration problems can impact security, such as (OWASP):

- Unpatched security flaws.
- Software flaws or misconfigurations that permit directory listing and directory traversal attacks.
- Unnecessary default, backup, including scripts, applications, and web pages.
- Improper file and directory permissions.

Secure Configuration Management

Software applications go through frequent cycles of updates, bug fixes, etc. From a security perspective and in comparison, with physical devices, IoT, etc. frequent software updates can facilitate quick vulnerabilities or security bug fixes, once discovered. On the other hand, software configuration management processes that have such frequent cycles of changes can be cumbersome. Without proper procedures to evaluate the need for bug fixes (e.g., impact analysis) and without proper procedures to fix those bugs and to document changes that occurred to the software, systems, and manuals, the purpose and justification to implement those fixes can be defeated.

In addition to bug fixes, there is a need to remove or disable un-needed or useless features so that they may not be used to create backdoors. There is a need for structured, period, and comprehensive security testing for the different installed software applications. Discovered vulnerabilities should be handled appropriately based on a structured software change and configuration management processes.

Security Controls and Policies to Support Secure Configuration Management

Vulnerabilities in software configuration are not uncommon. Security controls and policies should ensure the right procedures to plan for such problems.

Followings are examples of such security controls and policies:

- Use only verified software: Use versions of operating systems, web browsers, and applications that are known, verified, and vendor supported.
- Include proper policies for bug fixes, updates, and patches using proper procedures and tools.
- Conduct periodic and infrequent security testing and vulnerability assessment.
- Create and maintain software and hardware inventories.
- Implement a secure baseline for operating systems, DMBS, servers, etc.
- Disable unnecessary or unused software and hardware features, peripherals, user accounts, etc.
- Implement proper procedures related to access control such as "minimum privilege" and "layered and distributed access control."

S0090: Skill in Analyzing Anomalous Code as Malicious or Benign

Fig. 7.13 A sample scanning result for a popular Trojan

Malware Data Analysis Using Public Malware Scanners

There are many free malware scanners such as AutoShun, PhishLabs, Kaspersky, StopBadware, Sophos, or Netcraft. The testing output from the different malware scanners indicates that frequently they may have different decisions on the same files. It indicates that they also employ different detection techniques or methods as in Akour et al.

Figure 7.13 shows a sample scanning result for a popular Trojan. Although it is a popular Trojan, yet (1) 9 malware scanners indicate this file as "clean," and (2) only 25 of the 67 malware scanners that identify this file as malware, clearly indicate that its type is a Trojan. Malware scanners have to complete scanning systems with usually a large number of files. They have to analyze each subject or suspect file with static and possibly dynamic methods. Additionally, they have to be accurate and avoid different cases of false positives and negatives. Those are also other examples of challenges facing malware scanners and all data analysis activities related to malware analysis and detection.

Malware Clustering and Classification

In this experiment, we used a dataset collected from VirusTotal website to cluster malwares based on standard features. A sample below is extracted from RapidMiner, as a sample.

- Minimum four clusters, K-means as in Fig. 7.14.

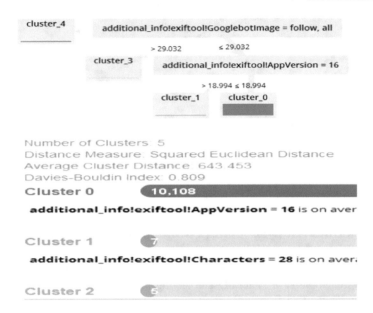

Fig. 7.14 K-means for minimum of four clusters

```
Country/Rank <= -1: FALSE (292.0/11.0)
Country/Rank > -1
|   Country/Rank <= 29965
|   |   token_count <= 3: TRUE (439.0/6.0)
|   |   token_count > 3
|   |   |   token_count <= 4
|   |   |   |   URL Rank <= 101719: TRUE (70.0/1.0)
|   |   |   |   URL Rank > 101719: FALSE (4.0/1.0)
|   |   |   token_count > 4: FALSE (5.0/1.0)
|   Country/Rank > 29965
|   |   No_of_dots <= 1
|   |   |   Length_of_url <= 15: FALSE (5.0)
|   |   |   Length_of_url > 15
|   |   |   |   Length_of_url <= 18: TRUE (8.0/1.0)
|   |   |   |   Length_of_url > 18
|   |   |   |   |   URL Rank <= 950724: TRUE (9.0/3.0)
|   |   |   |   |   URL Rank > 950724: FALSE (7.0)
|   |   No_of_dots > 1: FALSE (19.0)

Number of Leaves  :      10

Size of the tree :      19
```

Fig. 7.15 An example of classification extracted based on decision tree classification

```
Kappa statistic                    0.8705
Mean absolute error                0.0854
Root mean squared error            0.2333
Relative absolute error           18.0519 %
Root relative squared error       47.9721 %


=== Detailed Accuracy By Class ===

               TP Rate  FP Rate  Precision  Recall  F-Measure  MCC    ROC Area  PRC Area  Class
               0.939    0.063    0.960      0.939   0.949      0.871  0.946     0.951     TRUE
               0.938    0.061    0.905      0.938   0.921      0.871  0.946     0.920     FALSE
Weighted Avg.  0.938    0.062    0.939      0.938   0.939      0.871  0.946     0.939
```

Fig. 7.16 Classification performance/accuracy metrics

Malicious URL Links Classification

In this experiment, we collected a dataset list 453,340 links malicious and benign from different public sources. The links class labels for all links (e.g., whether malicious or benign) is verified manually and confirmed from different sources. About 25% of the links are malicious (109,258 links). We believe that the size of both groups is significant in the count and in comparison, to the other class label, which can help make the realistic classification.

One of the significant findings that we noticed in several classification algorithms is the importance of popularity metrics to indicate that links are not malicious. In other words, if links or their URLs are known and has ranks (i.e., not -1 in our formula), then most likely such links or URLs will not be malicious. We used rankings from Alexa (URL Rank) and SimilarWeb global rank (Country/Rank). The tree shows that malicious links are those links that either don't have a rank or have a high-rank value (i.e., less popular).

Figure 7.16 shows overall performance metrics for the results in Fig. 7.15, which shows significant or acceptable accuracy.

S0091: Skill in Analyzing Volatile Data.

Memory Analysis with Volatility

- Download the images file from https://www.memoryanalysis.net/amf. The folder has six Linux Bin sample memories. Pick two of those bin Linux memories. Then using Volatility memory analysis tool, make a comparison for each output from the tool (using the same commands described in https://samsclass.info/121/proj/ p4-Volatility.htm. In your comparison table focus on the differences in the output from the two images explaining in your own words for each case what is this difference and why the two images have such difference although they have the same operating system. Notice that changes can come from different hardware, software, settings, or users.

- In this part, we will learn basics to use Volatility framework for memory forensics. Volatility framework is prevalent. It is used in many security training courses. Several computer legal references indicate that the framework has been effectively used to discover exciting pieces of evidence in many forensic cases. Follow the steps described in the link (https://samsclass.info/121/proj/p4-Volatility.htm) to perform memory analysis. Rather than using the memory dump described in the link, dump your computer memory using Dumpit tool: http://qpdownload.com/dumpit/, or any dumping tool listed in (http://forensicswiki.org/wiki/Tools:Memory_Imaging). You can also use publicly available memory dumps (e.g., 1. https://www.memoryanalysis.net/amf) or (2. https://code.google.com/archive/p/volatility/wikis/SampleMemoryImages.wiki). One example for the second dataset is Stuxnet malware. This is a link describing how Volatility was used to analyze Stuxnet: https://volatility-labs.blogspot.com/2013/05/movp-ii-21-rsa-private-keys-and.html).

S0130: Skill in Writing Scripts Using R, Python, PIG, HIVE, SQL, etc.

Knowledge management is a structured and systematic process to extract learning from past activities to make better future decisions. Knowledge Management processes deliver measurable benefits. We will focus in this section on examples of using machine learning (ML) techniques in cyber operations, especially for cyber analysts.

A ML approach usually consists of two phases: training and testing. Often, the following steps are performed:

- Identify class attributes (features) and classes (class labels) from training data.
- Identify a subset of the attributes necessary for classification (i.e., dimensionality reduction, feature selection, etc.).
- Divide data into training and testing, learn the model using the training data.
- Use the trained model to classify the unknown data.

Some of the popular algorithms: ANN, SVM, GA, KNN, Random forest, HMM, etc.

Readers are expected to learn some of the popular data mining tools such as:

- Python: One of the most popular programming/scripting languages for cybersecurity and data analytics. Several open-source IDEs can be used to write and execute Python code such as PyCharm and Anaconda. Some of the popular Python libraries to learn in this scope: Scikit Learn and TensorFlow.

- R: Of the popular GUI IDEs based on R is R-Studio. Users can write scripts that utilize rich libraries built and available in R.
- Weka: A accessible but straightforward open-source GUI based data mining tool. Libraries also exist to export Weka to Java.
- Knime (knime.com), written in Java, Knime is a free and open-source data analytics' reporting and integration platform.
- RapidMiner.
- H_2O.
- MATLAB/Octave.
- Julia.
- Several tools and libraries in Java (e.g., see deep learning for Java: https://deeplearning4j.org).

Deep analysis in data science using different algorithms to look for deep or hidden knowledge, intelligence, or patterns in analyzed data. Some of the open-source tools and mechanisms that can be used in forensics deep analysis:

- Java packages such as DL4j (https://deeplearning4j.org) and TensorFlow: (https://www.tensorflow.org/install/install_java)
- Python with Deep learning libraries such as:

  - Caffe: http://caffe.berkeleyvision.org/
  - Theano: http://deeplearning.net/software/theano/
  - TensorFlow https://www.tensorflow.org/

- R and RStudio: Libraries such as TensorFlow and Keras (https://keras.io)

  MATLAB: https://www.mathworks.com/solutions/deep-learning.html.

References

1. Software Engineering Research Laboratory. (2008). Institute of Electrical and Electronics Engineers, Inc. www.swebok.org.
2. Alsmadi, I., & Dieri, M. (2009). Separation of concerns in teaching software engineering. In *Proceedings of CISSE 2009*. The USA. http://conference.cisse2009.org/proceedings.aspx.
3. Humphrey, W. (1989). *Managing the software process*. Reading, MA: Addison-Wesley.
4. Pressman, R. S. (1996). *A manager's guide to software engineering*. New York: McGraw-Hill.
5. Royce, W. (1970). Managing the development of large software systems. *Proceedings of IEEE WESCON, 26*(25), 1–9.
6. Bischofberger, W. R., & Pomberger, G. (1992). *Prototyping-oriented software development– concepts and tools*. Springer-Verlag.
7. Boehm, B. W. (1988). A spiral model of software development and enhancement. *IEEE Computer, 21*(5), 61–72.
8. Larman, C., & Basili, R. V. (2003). Iterative and incremental development: A brief history. *IEEE Computer, 36*(6), 47–56.
9. McConnell, S. (1995). *Rapid development*. Redmond, Washington, USA: Microsoft Press.
10. Booch, G., Maksimchuk, R., Engle, M., Young, B., Conallen, J., & Houston, K. (2007). *Object-oriented analysis and design with applications* (3rd ed.). Addison-Wesley Professional.

11. Kruchten, P. (2003). *The rational unified process: An introduction* (3rd ed.). Boston, MA, USA: Addison-Wesley Longman Publishing Co. Inc.
12. Boehm, B., & Turner, R. (2003). *Balancing agility and discipline: A guide for the perplexed.* Boston: Addison-Wesley.
13. Schuh, P. (2005). *Integrating agile development in the real world.* Hingham. MA: Charles River Media.
14. Schwaber, K., & Beedle, M. (2002). *Agile software development with scrum.* Upper Saddle River, NJ: Prentice-Hall.
15. Palmer, S. R., & Felsing, J. M. (2002). *A practical guide to feature-driven development.* Prentice-Hall International.
16. Paulk, M. (2001). Extreme programming from a CMM perspective. *IEEE Software, 2001,* 19–26.
17. Raymond, E. S. (1998). The cathedral and the bazaar. *First Monday, 3*(3).
18. Wayner, P.(2000) Free For All. HarperCollins, New York, 2000.
19. Sharma, S., Sugumaram, V., & Rajagopalan, B. (2002). A framework for creating hybrid-open source software communities. *Information Systems Journal, 12*(1), 7.
20. Madey, G., Freeh, V., & Tynan, R. (2002), The Open Source Software Development Phenomenon: An Analysis Based On Social Network Theory. AMCIS.

Chapter 8
System Administration

Izzat Alsmadi

K0066: Knowledge of Privacy Impact Assessments

US 2002 E-government Act identifies the need to conduct Privacy impact assessment (PIA) for government programs. Those assessments are usually conducted by federal agencies.

Federal Trade Commission (FTC: www.ftc.gov) access control policies require that all systems use access control mechanisms to restrict access to system functionality and sensitive data, including PII. Multi-factor authentication is required for certain privileged users. All government contractor personnel with access to the FTC's network are required to take the annual Information Security and Privacy Awareness Training. Profiles that are created by contractor personnel are audited regularly. Privacy incident response plans are also required to handle privacy incidents. Violation of such plans, contractors will be subject to the FTC's breach notification response plan.

According to FTC policies, in most instances, they collect minimal personal information, such as name, address, telephone number, or email address. In limited cases, depending on the nature of the case, other information such as Social Security numbers, account numbers, or mortgage or health information will be collected. FTC's Privacy Act rules including procedures, timelines, and instructions for submitting Privacy Act requests, and a list of FTC systems of records that are exempt from the Act's requirements are published at the electronic code of federal regulations: 16 C.F.R. 4.13 (https://www.ecfr.gov). A PIA form, in addition to an automatic workflow and streamlined review and approval process, has been developed for consistency and ease of use. The form and additional guidance about PIAs are available as an internal document for NCUA (ncua.gov).

A PIA is both an analysis and a formal document detailing the process and the outcome of the assessment or analysis. The privacy assessment is a practical method of

I. Alsmadi et al., *The NICE Cyber Security Framework*,
https://doi.org/10.1007/978-3-030-41987-5_8

evaluating privacy in information systems and collections and documented assurance that privacy issues have been identified and adequately addressed.

PIA techniques identify and assess privacy risks in systems and assets. The critical element is to identify human personal private information: Personally Identifiable Information (PII). The exact process should be put in place to explain how to collect, use, protect, share, and maintain PII. PIA model should be able to:

- Allow users to be aware of what is collected about them and enable them to consent to allow such collections or not. Users should also be aware of locations of cameras in buildings, logging procedures in web and other types of applications, and so related to any application with information collection nature. Users should be aware of such collection methods and should be able to decline self-collection under certain conditions.

Individuals may make a request under the Privacy Act for access to information-maintained bout themselves. Access to such information is subject to certain exemptions

- Judge whether collected PII complies with relevant regulations.
- Identify and assess risks associated with the collection and maintenance process of PII.
- Identify and assess methods to protect PII information at rest in databases, etc. and also in transition or communication.
- Procedures should be put in place to ensure that the information maintained is accurate, complete, and up-to-date. Users should be able to challenge and modify incorrect collected information about them.

A PIA should accomplish three goals [1]:

- Ensure conformance with applicable legal, regulatory, and policy requirements for privacy.
- Determine the risks and effects
- Evaluate protections and alternative processes to mitigate potential privacy risks.

A PIA should be completed when any of the following activities occur [2]:

1. Developing or procuring any new technologies or systems that handle or collect personal information.
2. Developing system revisions.
3. Initiating a new electronic collection of information in an identifiable form consistent with the Paperwork Reduction Act (PRA).
4. Issuing a new or updated rule-making that affects personal information.
5. Categorizing System Security Controls as a high or moderately major.

K0261: Knowledge of Payment Card Industry (PCI) data security standards
The Payment Card Industry (PCI) Data Security Standards (DSS) is a global information security standard that is designed to prevent fraud through increased control

of credit card data. Organizations must follow PCI DSS standards if they accept payment cards from major credit card brands (e.g., Visa, MasterCard, American Express, etc.).

PCI DSS applies to any organization that transmits, processes, or stores payment card transactions or cardholder information. The PCI standards consist of six critical objectives over 12 significant requirements.

Companies that are not PCI compliant can face fines and, in some cases, have their payment card privileges revoked. Regardless of how many transactions they complete, companies must demonstrate PCI compliance.

In addition to technical compliance, PCI compliance requires establishing and maintaining appropriate policies, procedures, and technology to ensure consistent compliance through continuous protection of collected, stored, or transmitted data.

PCI Compliance Solutions

PCI DSS was established by the major card brands. All businesses that process, store, or transmit payment cards data are required to implement the requirements outlined in the PCI DSS to prevent cardholder data theft. Such PCI compliance is required and verified periodically. Merchants are required to validate and report their compliance to their merchant processor.

The compliance solutions usually combine technical and vulnerability screening or testing, awareness programs, and also data breach protection mechanisms or security controls.

K0262: Knowledge of Personal Health Information (PHI) data security standards

Privacy rules address the appropriate safeguards that are required to protect the privacy of personal health information (PHI). They assign limits and conditions concerning the use and disclosure of personal information collected and stored by healthcare organizations or any other businesses affiliated with these organizations. In the US, this is regulated under the Health Insurance Portability and Accountability Act (HIPAA).

HIPAA regulations ensure that PHI remains protected, and there are significant personal and institutional fines for non-compliance to those regulations.

Health Breach Notification Rule: **Under Health Breach Notification Rule** healthcare organizations or any other businesses affiliated with these organizations are required to notify their customers, the Federal Trade Commission, and in some cases the media if there's a breach of that health information.

Health Information Technology for Economic and Clinical Health (HITECH) Act was a component of the American Recovery and Reinvestment Act (ARRA) of 2009. It was part of an effort of the federal government to support the widespread adoption of Electronic Health Records (HER).

HIPAA covers regulations or rules for privacy as well as security. HIPAA sets the baseline for sensitive patient data protection with its Privacy and security rules.

According to these rules, any organization that deals with PHI must have in place physical, network, and process security measures and follow them to ensure HIPAA compliance.

HIPAA security rules aim at protecting health information that is held or transferred in electronic forms. The security rules operationalize the Privacy Rule's protections by mapping out the technical and nontechnical safeguards that must be put in place by covered entities to protect individuals' personal health information.

HIPAA regulations cover specific requirements for data management:

- Personally Identifiable Information (PII) should be adequately preserved in the health organization and should only be visible to the right personnel. Health regulations identified 18 different personal identifiers (Name, Address, dates elements (including birth date, admission date, discharge date, date of death, and exact age if over 89), Telephone number Fax number, Email address, Social security number, Medical record number, health plan beneficiary number, Account number, Certificate or license number, Any vehicle or other device serial number, Web URL, Internet protocol (IP) address, Finger or voice print, Photographic image and any other characteristic that could uniquely identify the individual).
- Data can be shared outside the organization if all 18 identifiers are removed.
- Identifiable health information should never be stored with cloud storage providers or shared using web-based email as information stored in these environments may be considered the property of the cloud vendor or email provider.

HIPAA security rules require organizations to implement the following protections:

- Ensure the confidentiality, integrity, and availability (CIA) of all electronic personal or protected health information (e-PHI) that they create, receive, maintain, or transmit.
- Identify and protect against *reasonably* anticipated threats to the security or integrity of the e-PHI information.
- Protect against reasonably anticipated, impermissible uses or information disclosures.
- Ensure compliance to relevant regulations.

K0290: Knowledge of systems security testing and evaluation methods

Systems such as operating systems, database management systems, and the full range spectrum of information systems in the different domains can be subjected to different types of risks and threats. Security testing mechanisms in those systems must take into consideration all possible threats and risks that can be caused by unintentional or malicious intents. Unlike standalone applications, systems are typically distributed on different physical locations or assets. As such, many of the threats can come from non-physical access to those systems (i.e., through the network or the Internet).

Security testing at the system level validates that all security requirements are fulfilled. Testing starts with assuring that all specified security requirements or features exist. Testing then extends to ensure that the whole system as a whole works properly and resists all types of expected attacks.

Methods of formal validation, simple validation (e.g., reviewing, interviewing, discussion, validation, and system testing) and simulation of intrusion detection can be used to implement security testing.

Security testing is different from security evaluation. Security evaluation is typically performed by a third party or external authorized organizations or agents. On the other hand, they are interrelated in many aspects. For example, security testing can be viewed as part of the security evaluation, and both depend on similar security requirements of the target systems.

To plan sufficient and appropriate test cases for security testing, known security problems or bugs in similar systems should be studied. Standards and regulations should also be taken into consideration. Such security tests should be frequently visited as threats and security breaches continuously change and evolve. Security requirements of the target system should be analyzed and extracted based on security objectives, functional requirements and/or security mechanisms of the system and known testing and evaluation standards in the relevant field [3].

System-level security testing is essential but not enough by itself. Security testing by itself is ineffective in identifying or detecting security threats and weaknesses such as malicious codes, Trojans, backdoors, logic bombs, and other malware. System security testing should be viewed as one layer in the multi-layers overall security defense system.

Certain activities relevant to software security, such as stress testing, are often carried out at the system level. Penetration testing is also carried out at the system level. A vulnerability that can be exploited during system testing will be exploitable by attackers. Under resource constraints, these are the most critical vulnerabilities to fix, and they are also the ones that should be taken most seriously by developers. While defense in depth demands that individual executable system elements be secure, it also demands rigorous security at the system level to help prevent attackers from getting the first step in the first place.

K0291: Knowledge of the enterprise information technology (IT) architectural concepts and patterns (e.g., baseline, validated design, and target architectures.)
Enterprise Architecture (EA) is the analysis and design of an enterprise, including its current and future states from strategy, business, and technology perspectives. This helps to integrate and manage IT resources from strategic and business-driven viewpoints. It is the process of translating business vision and strategy into useful enterprise model by creating, communicating and improving the essential requirements, principles, and models that describe the enterprise's future state and enable its continuous evolution.

An enterprise information technology architecture (EITA) is a unified framework for managing and operating information systems and resources in an enterprise or organization-level. EAs are designed based on the following major principles:

- Focus on organization-level business needs and long-term strategic goals.
- Data and infrastructure as an investment with long-term values and benefits.
- System integration through sound design and use of applicable IT standards.

- Coordination, collaboration, and sharing of resources (e.g., data, systems, applications, support).

To implement technology at the enterprise level as efficiently and effectively as possible, it is necessary to view all systems/applications as a single enterprise made up of all entities which share the common goal for public services and the management of public resources rather than individual, autonomous applications or systems. The information technology architecture provides the building blocks needed to realize such enterprise-level architecture.

The design and deployment of the different components in the enterprise architecture should be evaluated and tested. Unlike testing in software application, evaluating the design does not include executing any application. Design verification or validation involves a combination of formal and informal techniques. Given a specific baseline to be tested, several compliance checks can be performed for each requirement, resulting in validation results, several compliance check results. "Review Baseline" represents the compliance assessment process.

In recent agile methodology, designs are validated with less time and formality as the process is evolutional; design may change in the next cycles, and also design can be revalidated in those next cycles. Capture just enough to understand the implications of transitioning to the target architecture, and no more.

The validated design may be determined for the product and its design and manufacturing processes. For example, Cisco Validated Designs (CVDs) describe systems and solutions that are developed, tested, and documented to facilitate faster, more reliable, and more predictable deployments of Cisco products. CVDs are provided in three formats: Design Guides, system assurance guides, and application deployment guides.

In an Enterprise Architecture, EA, a baseline includes several project artifacts that are reviewed and agreed on by their immediate stakeholders and system managers and which form the basis for further system or enterprise development. An EA baseline is essential for the evaluation of IT assets and future investments in them.

A baseline architecture provides the opportunity for design compliance assessment, as it describes the agreed-upon basis for the remainder of the project and yet allows for intervention in cases of non-compliance. In this way, deviations from the architecture can be identified, and steps should be put in place to correct them. Without a clear baseline, the evaluation process will be poorly grounded.

Baseline architectures can be challenging to justify higher management and stakeholders who have a more significant aim for target architectures. The importance of the baseline architectures is to establish the initial stage that will allow the transitions eventually to the target architectures. It is often the case that documentations and models will exist that can be mined to collect material from populating the baseline repository.

Target architectures are the focus of executives and managers because they define the architectures that will realize the business strategies and goals and deliver values to the enterprise. Once these target architectures are known, and there are enough details in the baseline architectures, the architecture team can set about the more difficult task of defining the transition architectures.

K0299: Knowledge in determining how a security system should work (including its resilience and dependability capabilities) and how changes in conditions, operations, or the environment will affect these outcomes.

Information systems face challenges from different categories, from technical and software failures to natural disasters, human mistakes, or malicious threats or attacks. Other challenges related to quality or non-functional requirements include: providing services with acceptable levels of performance, reliability, robustness, usability, etc.

MITRE [4] identified the following significant security design principles:

- Modularity and layering
- Defense in depth
- Least common mechanisms
- Security evolvability
- Isolation
- Least privilege.

Resilience is the ability of the systems to provide and maintain an acceptable level of services in the face of various faults and challenges to normal operations. Resilience and fault tolerance both address operations in the presence of faults or other challenges. Faults can arise for many reasons, including malicious actions, and techniques from fault tolerance may help to enhance resilience regardless of cause.

MITRE [4] identified the following major resilience design principles:

- Functional redundancy
- Layered defense
- Complexity avoidance
- Localized capacity
- Human backup
- Reorganization.

When considering resilience, the following should be observed:

- What is an acceptable level of service? When faults or security problems are occurring, system resilience aims to ensure the ability and continuity of providing acceptable levels of services in comparison with normal situations or operations. When such services are provided to other companies or partners, it is essential to defined an acceptable service level agreement or service (SLA, SLS).
- Which provisions can ensure the ability to provide and maintain a certain level of services?
- What faults and challenges our systems and services are facing?
- What is considered to be normal operations in our business functions?
- How to measure system risks and impacts and how to assess their consequences? Acceptable service level definitions can be refined based on further classification of the service disruption impact (SDA); the significance of the service impact

can be quantified using several impact metrics such as the extent of the impacted in terms of users, services or network portions or in terms of recovery times. Figure 8.1 shows an example of the difference between resilient (1) vs. non-resilient (2) services.

- How will risk be managed for the systems and networks services?

Organizations should manage such risks by:

- Determining the risk management significant objectives
- Reducing failure probabilities, the likelihood of faults and risks
- Reducing consequences from failures such as service disruption, financial, technical, etc. damages and other types of impacts.
- Reducing time to recovery (TTR): Time to recover to regular services and operations
- Deciding risk mitigations and tolerance: the level of risks an organization is willing to accept or deal with.

K0363: Knowledge of auditing and logging procedures (including server-based logging)

Most information systems include auditing or logging mechanisms to log a variety

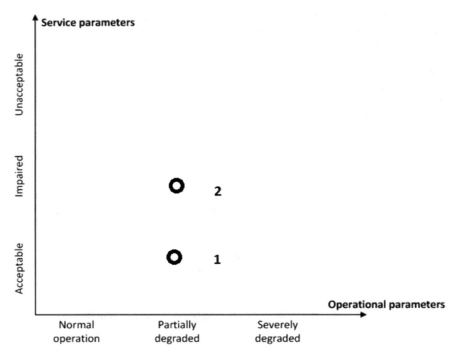

Fig. 8.1 Resilient versus non-resilient services

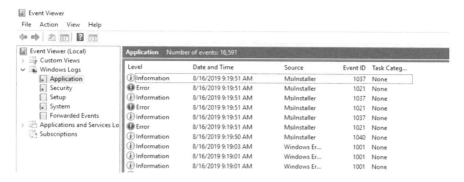

Fig. 8.2 An example of Windows applications event viewer logging

of different activities that occur throughout the system. We will cover examples of applications and systems and their logging and auditing processes.

Microsoft Windows Event Viewer as an Example of Operating Systems Logging

Event viewers in Windows operating systems include records and logs of many events that occur in the operating system and installed applications. While they meant mainly for maintenance and troubleshooting purposes, yet they can be precious for forensic investigations.

Event viewers include three significant categories of loggings: System (i.e., operating system logging activities), applications, and security loggings. Logged events in those categories can be in one of three types: Error, warning, or information. Each event includes information on when the event occurs, through which user, application, etc. Users can also create customized event logs in the applications they use or develop. Events can be exported as text files, and many tools can be used to extract artifacts from those events.

Logging can be for maintenance purposes (e.g., Windows Applications logging, information, warning and errors, Fig. 8.2). It can also be for security or forensic purposes (e.g., tracking system loggings, system resources' access, etc.).

Internet Usage Logging and Auditing

Using web servers/applications logs, systems log users Internet activities. Many regulations require certain levels of such loggings. Those can be used for troubleshooting of Internet problems, but most are used to track security breaches and also in digital forensic investigations.

The first essential types of applications related to Internet users that should be investigated are web browsers. Web browsers are software applications that enable users to access different websites through the Internet. History of those browsers can help us see which websites subject user was visiting, which pages on those websites when they were visiting those websites and much other relevant information. Of course, the value and critical of such information can vary from one forensic case to another.

From OSI layers' perspective, most of the information we can extract from web browsers can fall within layer 7 (i.e., application layer). In comparison to network analysis and forensic tools (e.g., Wireshark) that typically detect information in lower layers (e.g., L2–L3), information that can be extracted from browsers' history is easier to use and understand by humans or investigators without the need for significant skills or special high-end forensic tools. We will describe briefly how to search for Internet usage history in Windows most three popular browsers: MS Internet Explorer, Google Chrome, and Firefox.

• MS Internet explorer
Microsoft has its browser that will be installed with Windows operating systems. The browser options can be used to view or set the location where browsing history is stored.

Users can also make/change settings related to the size of the history or cache or for how long to keep such cache. Users can opt not to store browsing history. Internet browsing history can be used to accelerate bringing pages from the Internet where some pages can be brought from the cache if they exist without the need to collect them from the Internet. Users can also clear their browsing history. In such cases, forensic investigators can search for pointers for Internet history (e.g., cookies, recent files or shortcuts). Those can show indications of visited websites, watched videos, pictures, files, etc. while they may not help in retrieving those artifacts.

• Google Chrome
Google has two main versions of browsers: Google Chrome and Chromium. Chromium is recently introduced to target Desktop environments. There are some differences between the two versions on where the default location of their Internet browsing history. In most recent Windows operating systems, this default location is in the user application data folder (e.g., C:\Users\%USERNAME%\AppData\Local\Google\Chrome\User Data\Default\Preferences). Configuration can be seen or changed through access to Google Chrome configuration menus. In addition to cache or history content, the folder can contain many artifacts that can be useful for forensic investigations.

One of the main files to investigate in this folder is the (History file). This history file uses the SQLite database format.

Different time stamps can be extracted about the user's visits to the different websites, time, and duration of such visits.

Another important file in the same application data folder to investigate is the (Cookies) file. Similar to the (History) file, the Cookies file uses the SQLite database format.

• Mozilla Firefox
Firefox browser is a free and open-source browser developed by Mozilla foundation. Some of the distinguishing features in Firefox in comparison with other browsers are the support of a large number of add-ons, anonymous browsing, etc. Similar to other browsers, the default location of forensic artifacts in most recent Windows operating systems is the application data folder (e.g., C:\Users\%USERNAME%\AppData\Roaming\Mozilla\Firefox\Profiles\%PRO FILE%.default\places.sqlite).

Web Logs

If inspected machine or image has a web server (e.g., Microsoft Internet Information Server IIS, or Apache (https://www.apache.org/)), weblogs can be extracted from the machine. Such weblogs may include many forensic artifacts related to the users or visitors of this web site and their using activities or behaviors.

In digital forensics, there are several causes associated with hackers or employees accessing networks and sensitive information. However, we rarely see instances where organizations publicly announce employee had been investigated within the organization. Forensic investigations should be familiar with the Electronic Communications Privacy Act, CAN-SPAM Act, Communication Assistance, to Law Enforcement Act (CALEA), Foreign Intelligence Act (FISA), and USA Patriot Act. Equally important is the organization's willingness and responsibility to work with outside law enforcement depending on the degree of the activity. At this time, the organization's legal department will need to be notified to ensure that adequate and proper steps are taken.

Auditing and logging mechanisms can be used to search for evidence. Investigation teams should have the technical skills and the knowledge related to laws and regulations that make them capable of searching for, collecting, accurately handling, and using digital evidence. Due to the evolutionary nature of computing environments, digital-related laws evolved and continue to evolve rapidly. How much can valid and credible digital evidence be? Can we trust a weblog that traces a phishing attack to a particular user? Those are examples of open legal issues and concerns when it comes to digital investigations in general.

K0537: Knowledge of system administration concepts for the Unix/Linux and Windows operating systems (e.g., process management, directory structure, installed applications, Access Controls).
Details of system administration concepts can be found in details in different operating systems references. We will cover access control as one major module.

Access controls are considered as important security mechanisms. They usually target (authenticated users: Those users who can legally access subject information system or resource). This indicates that they typically come after an initial stage called (authentication). In authentication, the main goal is to decide whether a subject user, traffic, or request can be authenticated to access the information resource or not. As such, authentication security control decision or output is a binary of either, yes (authenticated; pass-in), or no (unauthenticated; block). Access control or authorization is then considered the second stage in this layered security control mechanism. For example, it is essential to decide whether the subject user has a view/read, modify, execute, etc. type of permission or privilege on subject information resource. In this chapter, we will cover issues related to access controls in operating systems, databases, websites, etc.

Operating systems evolve and continue to evolve with the continuous progress that we see in both the hardware and software industries. The operating system represents a sophisticated software application that is used to manage all other software applications installed on that same personal computer, server, etc. It is also responsible for controlling and managing the communication between the three main entities in an information system: Users, software applications, and hardware. The primary management modules that most mature operating systems include are memory, processes, file, disk, and network management. The tasks of access control in operating systems exist in different places and using different mechanisms. The basic structure depends on identifying users for the operating system and identifies their access levels on the different OS and system resources. In this sense, they combine access control with authentication. For example, users are prompted when they start an operating system to type a user name and password. Such a user name and password should exist in the directory of authenticated users in the OS with the right password. User names and passwords fall in the category of authentication mechanism (something you know).

Operating systems can also use other types of mechanisms such as something you have (e.g., an access card) or something you are (e.g., a fingerprint). System administrators can also decide different levels of constraints on the passwords that users are choosing for their accounts, how often they need to change it, etc. While user names and roles can be visible to other operating system users, passwords are encrypted. Rather than storing the passwords themselves, hash values of those passwords are stored in the (shadow) passwords' folder. Administrators (localhost administrators) have the highest possible permissions in OS resources. A user can be added as an administrator when they are included in the administrators' group.

In some operating systems, a particular user (root) is defined as (a superuser). Such a uniquely identified user may have special privileges that cannot be granted to other created users or accounts.

There are basic system access control principles that designers/administrators should consider:

- **Principle of Least Privilege**: This means that the default access is nothing for a user. Users will then be granted access to the resources that they need. In this regard, users are encouraged not to use administrator accounts all the time. If their accounts are exposed, with such high privileges, an attacker can significantly hurt the system. Alternatively, they should use standard or power user accounts and only elevate to administrators when they need to.

- **Separation of Duties**: Users should not accumulate responsibilities. They should not have open accounts that can play different roles on different occasions. Such roles should be divided.
- **Principle of Least Knowledge**: Similar to the principle of least privilege, users do not need to see resources that they have no associated tasks with. Intentionally or unintentionally users can abuse their privileges. In some phishing or social engineering types of attacks, those users can be victims, and their accounts can be used without their consent knowledge to commit attacks on systems or expose their resources.

Operating systems logically controls access through Access Control Lists (ACLs). ACLs are mechanisms or concrete implementations of access control models.

ACLs represent permissions on system objects to decide who can have view/create/modify/execute a system resource or object. In operating system ACLs, an access control entry (ACE) is configured using four parameters:

- A security identifier (SID)
- An access mask
- A flag about operations that can be performed on the object
- A flag to determine inherited permissions of the object.

K0608: Knowledge of Unix/Linux and Windows operating systems structures and internals (e.g., process management, directory structure, installed applications).

Details of system administration concepts can be found in detail in different operating systems references. We will cover the file systems as one primary module.

File systems represent the management modules of files/folders in operating systems. As such, the inherent most of their access control roles from the underlying operating system. File systems differentiate between user-generated file and operating system file. Access control decisions on operating system files are usually decided by the operating system access control. They may need special/administrative level permissions before users can change their attributes.

The concept of file/folder ownership evolves when operating systems start to allow more than one user to access/use the same operating system environment and applications. With this evolution also, operating systems now have the ability to audit files history to see who did what and when. Those are usually critical information components to know when conducting a computer or digital investigation.

In Microsoft and Windows operating systems, one of the main goals of moving from FAT to NTFS file systems was to enable features related to file/folder access control auditing. The FAT system was initially proposed before the era of operating system multi-users. In comparison with access control in database management systems, access control in file systems focuses on the file level access control. For a database, this will include, for example, roles on tables in comparison with roles on database schema in database access control.

UNIX file access permissions decide access on files based on three classes of users: Files' owners, members of the group which owns the file and all other users. Each of these three categories of users has permissions for reading, writing, and/or executing.

Windows File System

Windows File Allocation System (FAT) originally started in 1977 to be used on floppy disks. Eventually, it became the primary file system for disks in DOS and Windows operating systems. It has the three variants FAT12, FAT16, and FAT32. The number after (FAT) indicates the number of bits used for cluster addressing. FAT allocates logical disk spaces called "clusters" to files based on their sizes. Files are known in FAT by their names and the location of the clusters they are reserving. Each FAT system has a different cluster size in comparison with the physical size elements (i.e., sectors). FAT12 was proper for a floppy disk or small size disk storages.

Cluster size has a significant impact on disk utilization and performance. When cluster sizes are large, the disk will have more wasted space and less disk utilization. The size of a cluster can vary from one sector or 512 bytes to 128 sectors or 65 K bytes. Maximum volume size for FAT16 is 4 GB. This means that if you have a USB drive with a size of more than 4 GB, you can't format it as FAT16.

The majority of file systems in current Windows machines will be either FAT32 or NTFS. Any FAT file system will consist of the following three physical sections:

- A reserved area for file system information– This area starts from sector 0 or the first disk sector. This size of this area is given in the boot sector. The boot sector is the first sector in the reserved area.
- FAT area—For primary and backup FAT structures. One difference between FAT12/16 and FAT32 is that the root directory is fixed between the FAT and DATA areas in FAT12/16 while it can be anywhere in the data area in FAT32.
- Data area—Those are the clusters used for storing files and directory contents.

Hexadecimal editor tools can also be used to detect the file system type. The FAT file system has the values 0×55 and 0xAA in byte offsets 510 and 511 of the first sector.

Forensic investigators may have to deal with different challenges with data that can't be seen or extracted using standard searching mechanisms. Here are examples of those challenges:

- **Deleted data**: Suspects may try to delete specific files that they feel can be used as evidence against them. When files are deleted, their records in the FAT table are deleted. However, the actual file information is not erased unless if new content is added to the same ex-file location. Once files are deleted, and their FAT addresses are claimed as "empty addresses" new files can then use them.

The classic BTK killer case is an example of retrieving deleted files to be used as evidence. The suspect sent a floppy disk to officials in which they were able to retrieve his information from some deleted files.

- **Hidden data**: A hidden data or area in a disk is that data/area that is not seen by the file system. While most attempts in forensic cases can be related to either erasing or encrypting relevant data, however, it is possible that many occurrences of data hiding were not discovered. Here are different examples of possible hidden areas in a disk:

  – **Unused sectors** in the reserved area
  – **Slack spaces**: There are different types of slack spaces that can be used to hide data, including File, RAM, drive, etc. File systems used fixed-size containers called clusters to store files. The rest of the cluster at the end of the file is a file slack as this space cannot be used or claimed by other files. A file slack and the empty data between the last bit of file data to the end of the last cluster used by the file. Each file in the file system can have this "leftover," and the total disk space can be the total slack spaces from all files. The RAM slack happens in memory as data is written in memory in sectors (blocks of 512 bytes). As such, the last block in a retrieved file to the memory will be filled with random data to complete the last sector.

Slack spaces can be used from a legal perspective in two aspects:

- A professional hacker or suspect can craft a malicious file, or application to be hidden in some or all different partition or disk slacks. While this may seem to be complicated, however, it is not impossible. On the other hand, such acts will be hardened forensic investigators to detect.
- Slack spaces may keep data from earlier files. As a result, hackers may use tools to scan slack spaces looking for valuable information to steal. Users may assume that such data is deleted.

Data hiding using stenography and watermarking: Unlike encryption which disables readers from understanding data, while they can still read/see it, stenography and watermarking techniques are based on hiding some parts of data within other parts. Users with standard tools can see files or applications that they are supposed to see. On the other hand, other files or data portions are hidden within the visible data and can only be seen using specific methods and tools.

Every partition contains a boot sector. If the partition is not bootable, the boot sector in the partition is available to hide data.

Unlike old hard disks, modern hard disks can handle bad sectors themselves by remapping bad sectors to spare sectors. Clusters marked as bad may be used to hide data.

Unallocated/formatted data: Simple operating system tools can't see or extract data from a formatted disk or when some or all parts of the disk size are in "unallocated space." Many forensic tools can extract data from unallocated spaces. However, disk repartitions can cause data or parts of the data to be corrupted or destroyed.

Linux File System

Extended File System (Ext) is a family of file systems introduced with Linux and consists of Ext, Ext2, Ext3, Ext4. Ext file systems share many properties with NTFS: Access Control Lists (ACLs), fragments, undeletion and compression and journaling (beyond Ext2). An Ext file system starts with a superblock located at byte offset 1024 from the volume start (Block 0 or 1).

The first version (Ext) came to solve issues with Minix or Mint previous file systems. The two main issues Ext targets were extending maximum partition size up to 4 TB (Ext has a maximum partition size of 2 GB) and also the file name length to 14 characters. Soon Ext2 replaces Ext due to many problems in Ext file system, and Ext2 became the standard for Linux operating systems. The main problem with Ext2, which leads to introducing Ext3 is that Ext2 does not support journaling. As it has no journaling, Ext2 is still preferred on USB flash drives as it requires fewer write operations.

In Ext file systems, every file or directory is represented by an inode (index node). The inode includes metadata about the size, permission, ownership, and location on the physical disk of the file or directory, number of blocks used, access time, change time, modification time, deletion time, number of links, fragments. For forensic analysis, although they don't hold contents, inodes are the primary repository of metadata.

Understanding the journaling process in file systems can help forensic investigators recover deleted or corrupted files if necessary. Journaling is a process used in file systems for crash recovery situations. File system archives activities in typical situations so that when a problem occurs, it can recover to a safe situation. The journal process works by caching some or all of the data writes in a reserved part of the disk before they are committed to the file system.

In Ext3, the journal is a fixed-size log which regularly overwrites itself. The journaling process can be accomplished using different approaches or algorithms (e.g., Journal, ordered, and write-back) in the different operating systems. The first factor that impacts the journaling process is the performance or how much impact the journaling process can cause the operating system and routine operations. This takes the form of efficiency or resources' consumption as well as speed. The second factor that impacts or decides the journaling process is the amount of data to journal. There is a clear trade-off between performance and amount of data or metadata to log. From a forensic point of view, journal mode will be better than ordered or write-back as it offers metadata and recovery option, unlike the other two options that can only help in recovery (without significant metadata). The default option in Ext3 is "ordered," and hence little metadata can be extracted from investigated activities.

The journaling process will write/modify data in disks. Hence, it's essential for a disk forensic investigation to prevent this process while a disk is under investigation. For example, if a disk with EXT3 or EXT4 file system is hashed, then mounted/un-mounted then hashed again, this may create different hash values.

File Recovery in Ext File Systems

Similar to most file systems in Ext file systems, deleting a file does not erase file content but only the file system link or record of the file. For the file system, once a file is deleted, its space will be claimed unallocated, and any new system activity (e.g., creating new files, installing applications, operating system updates, or journaling) can use this unclaimed space.

If files are deleted and their space is not yet used by other files or applications, the process of retrieving such data or files is relatively easy and can be provided by a large number of tools from operating systems, forensic tools, free, open-source, etc. However, the process can be very complicated if this space is used partially or entirely by other files or applications.

Followings are steps that can be followed in a forensic investigation to recover a file or some files:

- Identify their inode numbers. In Ext, file systems files are uniquely linked with different inode numbers. Several tools can be tried to extract Ext file system inode numbers (e.g., TSK ils, ls, stat, istat, df, find, etc. The inodes do not contain the content of the files but their metadata.
- Using commands such as stat, istat, or find, you can make queries on specific inodes to get their status and metadata. (e.g., find. -inum 434404 –print).
- Knowing in which (block group) the subject inode exists, can help us search for more details about the block that the inode belongs to. For example, we can use a command such as fsstat to make queries on block groups. Block groups are described in a block or set of blocks called "group descriptor table."
- From the journal (if supported), inodes can be extracted using commands such as dumpe2fs, jcat, dcat, blkcat, etc.

Apple File System

Hierarchical file systems (HFS and HFS + or extended HFS) are file systems used by Apple in their operating systems; MAC as successors to earlier MAC file system; MFS. HFS replaces MFS to solve problems related with the evolution of disk sizes and the need to deal with large disk sizes, file sizes and names, etc. Similar to NTFS, HFS replaced MFS flat table structure with the Catalog File which uses a B-tree structure that could be searched very quickly even when size is large. HFS + is introduced in MAC OS 8.1 with better performance and internationalization/encoding features in comparison with HFS.

The Trash is represented on the file system as a hidden folder. Trash, on the root directory of the file system. However, similar to Windows, Trash folder content can be retrieved using the mouse or command line. HFS + preserves dates and time stamps when moving files to and from the Trash. Similar to other file systems, deleted files can be retrieved if their place is not overwritten.

K0609: Knowledge of virtual machine technologies.
K0610: Knowledge of virtualization products (VMware, Virtual PC).

Virtual or sandboxing environments allow users to run analysis machines in isolated environments from the hosting operating system. This can serve several goals, including:

- **Ready-setups**: Users may want to evaluate a tool that runs in a different operating system than their own. Tool setups can be very complex; they may need to experiment with those infrequently. They can have those experimental tools or environments available on external drives. They can run them on their machines whenever they want without actually impacting their host operating systems. Having such setups ready in virtual images help carry such setups and install them in different locations.
- **Integrity issues**: System administrators experimenting certain setups or troubleshooting specific problems or forensic Investigators analyzing a suspect malware should isolate analysis environments from their computing environments to (1) eliminate their possible tampering and hence risk evidence integrity, forensic investigators, or (2): isolate tested environments from operational environment, system administrators.
- **Security issues**: Analyzing malwares is typically conducting in sandboxed isolated environments. This ensures that executing or deploying those malwares will not cause them to infect testing environment or propagate further.

Some of the popular virtual environments include:

- VMware products (e.g., vSphere: https://www.vmware.com/products/vsphere. html, ESXi: https://www.vmware.com/products/esxi-and-esx.html and VMware Workstation: https://www.vmware.com/products/workstation-pro.html)
- Microsoft Hyper-V (https://docs.microsoft.com/en-us/virtualization/hyper-v-on-windows)
- Amazon Elastic Compute Cloud (EC2): aws.amazon.com/ec2
- Oracle VirtualBox (https://www.virtualbox.org/).
- CITRIX XenServer: https://xenserver.org.

References

1. DHS, Privacy Impact Assessments. https://www.dhs.gov/privacy-impact-assessments. Accessed August 2019.
2. Privacy Office, Office of Information Technology, Privacy impact assessment guide. https://www.sec.gov/about/privacy/piaguide.pdf. Revised January 2007.
3. Zhai, G., Niu, H., Yang, N., Tian, M., Liu, C., & Yang, H. (2009, December 10–12). Security testing for operating system and its system calls, from book Security technology. In International Conference, SecTech 2009, held as part of the Future Generation Information Technology Conference, FGIT 2009 (pp. 116–123). Jeju Island, Korea.
4. MITRE. (2017). *Deborah Bodeau and Richard Graubart* (p. 2017). MITRE January: Cyber Resiliency Design Principles.

Chapter 9
System Architecture

Lo'ai Tawalbeh

Introduction

To understand system architecture, it is vital to understand each component, how these two components work together, and how that relates to the information technology (IT) sector. Architecture is generally described as the science of building something. In fact, the Britannica dictionary defines architecture as, "The art and technique of designing and building, as distinguished from the skills associated with construction" [1]. Systems are usually a group of items designed to work together and can be objects across multiple layers. For example, humans interacting with hardware or software. From the information gathered previously of the two separate components, we can conclude that system architecture is the science of building a group of items designed to work together and serve a common purpose. Before we cover the architecture design itself, we will discuss the thinking process behind implementing systems and what information is considered when designing the complexity of these architectures. Since the scope of system architecture is broad, we will focus on network architecture specifically and cover Peer-to-Peer architecture in comparison to client/server architecture. Once we can identify the differences of these architectures and each of their benefits, we can fully grasp the design and innovation between the two types of architectures.

Why is system architecture important? A good IT system architecture provides a basis for all interested parties in a business to have a common understanding and communication of the system. IT system architectures can be very large and complex spanning several different departments and levels. The system architecture can provide a well-designed layout and a model to how the system works, how it communicates with other devices, and where these devices communicate. Where the devices communicate can have several different meanings, usually never meaning physically. When we refer to where we mean logically and at what point of the data transmission do these devices and objects interact with each other. In the ece.ubc.ca journal, the author writes, "Not only does architecture prescribe the structure of the

I. Alsmadi et al., *The NICE Cyber Security Framework*, https://doi.org/10.1007/978-3-030-41987-5_9

system being developed, but that structure becomes engraved in the structure of the development project" [2]. This is also an asset to developing a system architecture.

System Architecture Design

Network architectures each have a unique purpose to fulfill dependent on the organization and its expectations. Several variables exist while determining the structure and purpose of the system to design, Things that organizations consider before designing a system can include, cost, purpose, clients, lifecycle, and failover. This list is short in comparison to the entire overview used by an organization to finalize their design. When determining cost-effectiveness, we must consider the cost of the risk and cost of the data being protected. Cost can be monetary value, but cost can also be placed on non-tangible items like data and sensitive information or proprietary data. Purpose is significant because it identifies your audience and what you need to accomplish with target audience. For example, are we designing a system for internal resources only, or will the system need to accommodate outside clients and their data as well? This ties into identifying your client and the risk associated with 3rd party data. If our system must store another organization's sensitive information, then we must accommodate that and allow for additional spending and cost to ensure we have enough security and failover protection. All these factors will dictate the architecture and the scope of your system. Ultimately, understanding the factors and needs that are required, will help us design the best fit system architecture. The process of determining all these variables is called system scope. We will discuss system architecture tools and scope to understand the thought and design process that goes into creating a well-designed system [3].

Architecture Framework

We will not go into depth discussing architecture framework, but it essential to understand its importance within the system architecture and its distinction. For a network architecture to be functional, it uses architecture framework to ensure compatibility across different platforms and clients. This framework provides guidelines that will be consistent for the entire system. These frameworks enable entire systems to form a cohesion across all organizational requirements. The architecture framework addresses the concerns of all departments and parties and transitions those concerns into the design of the system architecture. For example, if our internal design requires the use of Windows servers and Cisco devices, then the framework is provisioned to have these devices to all work together fluidly. Subsequently, if an additional third-party only uses Mac devices, then the framework must include guidelines to ensure all future design and scalability be compatible with Mac devices as well. This, of course, is a simple scenario; the framework is much more complex and includes hundreds of adaptations and protocol requirements [4].

Scope of Architecture

The scope concept can be applied to networks and system design to assist in system architecture implementation. The scope of network architecture sets initial rules and establishes a working model for the system. Scopes can include any objects that will interact with the system inside or outside of the network design, data that will move between the system and its components, functions that the system will run, and configuration models for logical, physical, and virtual connections. When we mention objects, this can be hardware, software, or humans. We will review the three main scopes to understand how establishing a scope can solidify a strong technology system design.

Time Scope

- Time scope identifies organizational time constraints and timelines. A time scope will identify time boundaries and any previously scheduled projects and ensure the architecture is designed either in conjunction with other projects or at optimal intervals. A time scope will also set milestones or completion dates to ensure the network architecture is on schedule.

Detail Scope

- The detailed scope sets the boundaries for architecture detail and accessibility to all personnel and parties. This scope breaks down each level of access in the system to meet organizational requirements and system requirements. The detailed scope plays an important role when deciding how to balance the level of detail in a network to comply with all other business requirements. If the architecture detail is low, the system may encounter issues when meeting business needs, scalability, and availability. This can include easy access to devices from all personnel. If the architecture detail is too high, an implementation may be difficult and network maintenance times can increase due to complexity.

Organizational Scope

- The organizational scope determines how the network architecture will be designed per department and client. Different areas of the organization require different needs and not all should have identical architecture. This reasoning ties into cost-effectiveness and the need to only apply resources where they are needed. If in the future this should change, a well-designed network should easily allow for scalability and failover [5].

Network Architecture Design

Once the framework and scopes are determined for the network architecture, we can properly allocate resources to build our network. The infrastructure design includes all objects, logical, physical, and virtual applications. Some simple things to consider when implementing your design and scope. Accessibility is an essential criterion to consider, we want to establish how we will allow access and who we will allow access to. In addition, adopting a solid accessibility model will ensure all future adaptations and expansions will provide little error. If costs allow, avoid single points of failure wherever possible. Identify if there are any points that would create a failing reaction and adjust your system architecture to eliminate all points at risk. Assuming this doesn't interfere with organizational scopes and cost projections for the development of the infrastructure, vital areas of your system should contain multiple failover points to combat unforeseen threats as can be seen from Fig. 9.1.

Now that we have covered the fundamentals and steps leading up to the implementation of a network architecture, we will cover two specific types of network architecture. Since network architecture is a vast topic and ranges with hundreds of different design options, for this presentation we will discuss Peer-to-peer (P2P) architecture and Client/server architecture. These two are the main considerations when building a network architecture and serve different purposes. These architectures will be the focus of our project and we will identify the history, composition, and advantages/disadvantages of designing each system [6].

Peer-to-Peer Architecture

Peer-to-peer architecture has produced many protocols and applications through its innovation and design. In fact, P2P originated with the internet. In 1969 ARPANET was designed by a group of engineers and computer science professionals and connected UCLA, Stanford Research Institute, UC Santa Barbara, and University of Utah. This setup was not in a client/server format because it treated all the peers equally.

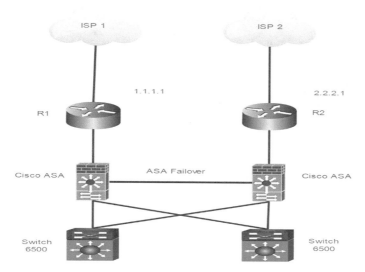

Fig. 9.1 Avoiding single point failover by multi-pathing

Peer-to-peer architecture also known as P2P is a unique design that shares all resources between computers. Meaning, there is not a centralized server that provides the workload for the entire system. Instead, all computers also known as nodes share data across the network equally, this design is known as Peer-to-peer. When one of the nodes needs to find data, it sends out a query and floods it across the network or nodes. Once the data is located from another node or nodes, the information is sent to the original requesting node. The popularity of Peer-to-peer architecture derives from its decentralized nature. The system does not rely on one main component to retrieve data or store data and therefore also reduces resources required to store and query large amounts of data. The architecture is easily scalable and provides for nodes to remain completely anonymous. This decentralized storage and anonymous nature allow for the flourishment of block-chain technology, which we will discuss in a later section. Figure 9.2 shows a standard P2P architecture design [7].

When a new node is connected to the network, it connects with any participating nodes that request or provide data. Once this link is formed it continues to grow as more nodes are discovered by the newly connected node. For example, if one peer knows the location of another peer within the network, they form what's called a direct edge to one another. Examples of popular Peer-to-peer architecture software includes; Bitcoin, Gnutella, BitTorrent, and OpenFT.

A P2P architecture can be classified into two categories, unstructured and structured networks. Unstructured Peer-to-peer architecture forms new links randomly and can be constructed easily as new nodes join the network. As previously mentioned, to find data, these nodes flood the network with queries to locate as many peers as possible. Once peers are located, new nodes copy these existing links and

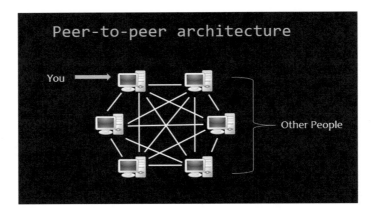

Fig. 9.2 Peer to peer architecture

form new links over time. There are some disadvantages to consider when using an unstructured P2P design. When a node floods the network with a query, this flood causes an increase in traffic and can impact data transfer speeds and connectivity for nodes. In addition, queries are not always resolved. IF a node requests data that is popular then it is likely that one of the nodes in the network will contain this data. However, if data is requested that is less sought after and available, the query will not be resolved.

Structured Peer-to-peer architecture overcomes the restrictions of unstructured P2P by maintaining a Distributed Hash Table (DHT). This table uses a hash function to assign value to the content and every peer in the network. The DHT allows peers to be responsible for specific parts of the content within the network by using a global protocol. This global protocol determines the peers and directs the query towards the selected content. The table can maintain a hash of preferred routes and knowledge of what nodes contain the section of data requested.

Client/Server Architecture

Client/server architecture is the more traditional setup you will see if organizations. This architecture design consists of clients or workstations that pull data from a centralized server. This design varies greatly from organization to organization, but the minimum architecture design must include at least one main server that carries the workload of resources that other clients pull from. A client is a typical computer that must rely on the server for its data and software. Due to this architecture, the server must carry the brunt of the workload while answering queries from its workstation. This type of architecture is called a centralized design as opposed to P2P architecture which we discussed early which relies on a decentralized design. To

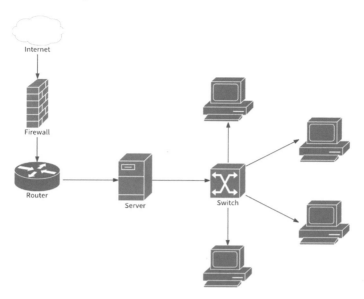

Fig. 9.3 Client server architecture

classify Client/server architecture, we can break it down into single-tier or multi-tier architecture designs. Figure 9.3 shows a simple client/server architecture.

Single-Tier Architecture client/server architecture is ideal for low traffic environments or system testing. A single tiered design puts all components required by the network on one server. In comparison, this is like having all the data on one node in a P2P design and is very demanding on the computer. However, servers are usually very expensive computers with powerful CPU's and large memory capacities to carry such workloads. In these single-tier designs, the one server carries all backend and front-end products, and everything is queried from the server. This design is straightforward and makes for easy management and maintenance.

Multi-tiered architecture is designed as a centralized network as well but contains multiple servers to handle unique resources. This divides the workload needed and assigns servers to execute only specific resources. For example, having a web server, database server, and FTP server. This design improves performance and increases data security. We should also mention that multi-tier architecture designs allow for failover protection which is recommended in all organizations that require data availability 24/7.

On the Cloud or On-Premises

On-Premise

The term on-premises refers to having a physical piece of hardware that is either owned by you or by a company. To give an example of this let's look at your computer. Whether you are on a desktop or laptop that piece of hardware is yours and can handle many things. These things include computing calculations, running programs, storing pictures and videos, as well as giving you access to the web. Sometimes a device like this can hold someone's entire life on the machine. With it requires a decent amount of power to handle all these tasks, but it does it pretty well. However, with a device doing so much for you there is a lot to lose. If your device were to get lost or stolen, you would lose a lot of data and important information that you may never get back again.

Companies will also have the same type of hardware located on-site but with much more performance. Instead of just a desktop computer like that one that you have been using at home, many companies use what are called servers. Servers are typically high-performance computers that are generally used for each task the company needs. One example might be a server used for holding all the files that you have access to when you are at work. Another might be tied to a program that will compute and process the information for you giving you a result. Just like your single desktop at home, you do face the risk of these servers failing whether it is from natural causes or other various risks [8].

On the Cloud

The cloud is not new and has just been renamed for the purpose of marketing. When people say they are afraid of cloud technology they do not realize that we have been using this technology for many years. One example that I like to give is email. Whether you use Gmail, Yahoo, or AOL they all have one thing in common and it is that those sites are the ones holding onto the emails. Emails get sent back and forth but when you are logging into a site to check your email what you are really doing it accessing a server owned by another company to retrieve those messages. That is why even if you are on a different computer you still login to the same site and see your messages just as they were before [9].

Now the cloud isn't a magical place where your information floats but rather it is a server that is not yours or held at your location. We will look at this in a way of storage. Many companies have emerged in the past several years just for the sole purpose of keeping your always accessible with an internet connection. One company we will investigate is Dropbox. Dropbox is known as a cloud storage company in which anybody with internet access can sign up for an account. After creating an account, you can send things such as pictures, videos, and files to be held by the company should you need to access them at any time. The best way to think this is like having a flash drive that you access over the internet. Now there is a limit on how much you can store in Dropbox but that limit can be increased for a monthly fee. The upside to this is no matter where you are you have access to the files you placed on the site if you have internet and a device that can access the website or application.

Within the past few years, we have done something even greater and turned the cloud into a machine that can perform tasks and processes much like a desktop computer or server could. The biggest thing out right now is what is called cloud computing. Without getting technical the best way to think about it is like you assigning somebody else a task. Sure, you can bake a cake at home but that will require you using your stove, ingredients, as well as time. However, depending on what you need it might be worth it to call your local bakery and have them make the cake for you. The same can be said for cloud computing where instead of having to use servers on-site you can delegate the task to other companies and just gather the results [10].

Now there are many cloud companies that will offer it all to you. The ones we will briefly discuss are AWS and Azure. AWS stands for Amazon Web Services and have been built by Amazon who have come a long way from just selling books and Azure is owned by Microsoft the same creators of the Windows operating systems. What these companies do is rent out their high-performance servers for individuals or companies to use for whatever application needed. Whether it is just to store documents, databases, or even process complex these companies act as the backbone for others big or small. (On-Premise).

Now that we have a better understanding of on-premises vs. the cloud, we can compare them in different aspects and why one might be better than the other. Now in no way is this report trying to tell you which one you should use but rather what may work best for you and your company and how it has helped others. We will discuss what is needed to run and maintain each and how many organizations worldwide prefer to use them. With more companies becoming digital, it is important to choose the right one not only as a cost saving but as a way that not only fulfills their needs but also their client's needs. While we did discuss cloud storage for the sake of argument, we will compare both as if being used for computing purposes.

Deployment

When looking into on-premises servers' deployment is a very long process. Just like cooking a 5-course meal you need somebody with experience and knowledge of how to do so. In this instance you need to have staff that can put the infrastructure together such as loading the servers with the correct operating system, setting the permission, creating documentation and more. You also must have staff that are readily available to be ready to service and maintain the servers that you have on-site should anything go wrong and no longer function. Not only are you paying for the hardware, but you are paying the salaries of workers who know that system. This process can also take several days or even weeks depending on what the function will be.

When looking at cloud computing the process to deploy a working environment is much quicker. AWS states that you can have a virtual cloud server ready within 15 min (Kellogg Company) Rather than taking days to have up and running you can have a server hosted by Microsoft ready to go in less than the time it takes for you to get ready. Because these cloud companies as hosting your virtual servers they are easier to maintain and do not need an expert and they are all tuned and created for the application purpose you need [11].

Cost

When it comes to money companies will find any way to be able to save a dollar on anything they do. When you look at on-premises hard not only do you have to pay for the actual server equipment you also have to pay for the software and license to load onto the server. The cost of servers isn't too much money but what many forget is that they are power hungry so you better make sure that you can handle the extra money they are taking each day. Not only are servers' power hungry but they can be large and most of the time you will need many to handle all your applications and task which leads you to needing space to be able to store them. With that all, you must also remember that they are running 24/7 and need to be constantly cooled so that they don't overheat causing damage to the equipment.

When you are looking at using a cloud provider you are paying them based on your usage. Because you are using their servers you no longer must worry about having a location to put them as well as keeping them cool. They are away from your environment so if you have an internet connection, they are available to you at any time of the day. After cost benefit is that you only pay for what you use most of the time. Because of this, you can easily estimate how much you are paying each month to utilize these cloud providers.

Control

Control is an important aspect where sometimes on-premises makes more sense in some aspect. Because your servers are on-site you have more reliability when it comes to the information they hold. There have been times when a cloud provider has been unavailable for a few minutes where data or processing power becomes unavailable which can cost companies a lot of money if they can't access information. With on-site servers, if you are on the same network as the equipment you can easily access what you need even if the internet is down. Another reason on-site helps in with highly regulated industries who can't risk a third-party handling information [12].

Now there is an upside to having cloud computing available and the control it gives. In the same scenario if you are somewhere else in the world and seem to be having issues a technician can investigate the system with the consoles that are provided from these cloud host machines. They can see what it is going on and should they need to make the appropriate changes over the internet as well as seek support from the company they are renting the equipment from. Another advantage is because you have more control you can scale up or down as needed. Using the management consoles given you can easily allocate more space or power to an individual server for the time being.

Summary

While Peer-to-Peer architectures seem like the clear choice for resource management and cost, there are some challenges that Peer-to-Peer architectures face. In situations where data is queried and must be retrieved, the client/server network holds a clear advantage. All data is centrally stored, and any query will retrieve the data being requested. In addition, all data can be easily copied and backed in case of any potential threats. In a P2P design, data is saved in pieces across all nodes, making it difficult to backup files. With the added failover and data security comes disadvantages as well as. Client/server architecture requires a much higher initial investment to design and continuous maintenance. These powerful servers are expensive and require top components and high capacity memory. To add to these costs, organizations must hire dedicated network administrators to manage all the resources. One way to make sense of the difference is client/server networks share network demands but not resources. While P2P networks share all the resources with each node providing its processing power and bandwidth to fulfill requests. If new nodes join the P2P network, the network increases its power by adding the resources of the new node. In contrast, when new workstations join a client/server architecture, the servers must share its resources with the new workstation plus the additionally installed workstations and this leads to decreased resources.

On the other hand, and given the increase in internet traffic and companies trying to stay ahead of the curve cloud computing is increasing not only with users taking advantage but also with the services that are being provided. There are times when having equipment on-site makes sense but in the digital age why have equipment that needs to be taken down for maintenance or becomes outdated after a year when you can have something that is also up to date. As you have read there are many pros and cons to moving your operations to the cloud or having them stay on-site but that can only be decided by you and your companies needs

References

1. Ackerman, J. S., Scruton, R., Collins, P., & Gowans, A. (2018, August 23). *Architecture*. Retrieved August 9, 2019, from https://www.britannica.com/topic/architecture.
2. Architectural Frameworks, Models, and Views. (2015, April 10). Retrieved August 10, 2019, from https://www.mitre.org/publications/systems-engineering-guide/se-lifecycle-building-blocks/system-architecture/architectural-frameworks-models-and-views.
3. Bursell, M. (2017, October 13). *5 traits of good systems architecture*. Retrieved August 10, 2019, from https://opensource.com/article/17/10/systems-architect.
4. Benkhelifa, E., Thomas, B. E., Tawalbeh, L., & Jararweh, Y. (2018). A framework and a process for digital forensic analysis on smart phones with multiple data logs. *International Journal of Embedded Systems, 10*(4), 323–333.
5. Acheson, P., & Dagli, C. (2016). Modeling resilience in system of systems architecture. *Procedia Computer Science, 95*, 111–118.
6. Dickerson, C., & Mavris, D. N. (2016). *Architecture and principles of systems engineering*. CRC Press.
7. Lo'ai, A. T., & Saldamli, G. (2019). Reconsidering big data security and privacy in cloud and mobile cloud systems. *Journal of King Saud University—Computer and Information Sciences*.
8. The Kellogg Company Case Study—Amazon Web Services (AWS). *Amazon*. aws.amazon.com/solutions/case-studies/kellogg-company/.
9. Jararweh, Y., Al-Ayyoub, M., & Song, H. (2017). Software-defined systems support for secure cloud computing based on data classification. *Annals of Telecommunications, 72*(5–6), 335–345.
10. Nori, A. K., Shukla, D., Christensen, Y., Krishnaprasad, M., & Sedukhin, I. (2016, May 10). Middleware services framework for on-premises and cloud deployment. U.S. Patent 9,336,060.
11. Lo'ai, A. T., & Habeeb, S. (2018). An integrated cloud based healthcare system. In *2018 Fifth International Conference on Internet of Things: Systems, Management and Security* (pp. 268–273). IEEE.
12. Habeeb, S., & Lo'ai, A. T. (2018). Feasibility study and requirements for mobile cloud healthcare systems in Saudi Arabia. In *2018 Third International Conference on Fog and Mobile Edge Computing (FMEC)* (pp. 300–304). IEEE.

Chapter 10
Threat Analysis

Chuck Easttom

Introduction

The NICE framework places emphasis on threat modeling. You cannot possibly devote maximal resources to all possible threats. Appropriate allocation of cybersecurity resources depends on identifying likely threats. Threat modeling goes beyond simply identifying threats, and includes modeling how a threat actor might execute a specific threat against a particular target. Fully exploring threat scenarios is an important process.

Threat Modeling

Identifying threats is only part of the process of assessing security. Clearly, any threat is a possible threat. However, modeling what threats are likely and their specific impact on a given organization is more complex. There are a few models based on acronyms to make them easier to remember. These are an excellent place to start.

STRIDE

STRIDE: Spoofing, Tampering, Repudiation, Information Disclosure, Denial of Service, Elevation of Privilege STRIDE was developed by Microsoft for identifying security threats in six separate categories. The threats are the letters in the acronym. The concept of using this tool is to ensure that you are modeling all of the enumerated threats.

© The Editor(s) (if applicable) and The Author(s), under exclusive license to Springer   207
Nature Switzerland AG 2020
I. Alsmadi et al., *The NICE Cyber Security Framework*,
https://doi.org/10.1007/978-3-030-41987-5_10

For example, if one wanted to model threats to an e-commerce website, each of the elements of STRIDE could be used. What methods might an attacker use to spoof a legitimate user? How might an attacker tamper with data? Is it possible for an attacker (insider or external) to deny their activity? What information might be disclosed and how? What are the vulnerabilities to Denial of Service? How might an attacker elevate privileges? All of these questions are driven by the STRIDE model and encompass the various attack vectors the e-commerce site should defend against.

DREAD

DREAD Damage Potential, Reproducibility, Exploitability, Affected Users, Discoverability. DREAD is a mnemonic for risk rating using five categories. How much damage would an attack cause? How easy is it for an attacker to reproduce this attack? How much effort is required to execute the attack? How many users will be impacted? And finally, how easy is it to discover the threat.

DREAD examines threats from a different perspective. It asks what the likelihood of an attack is and what damage it would cause. DREAD is an effective model for considering where to allocate resources. What attacks warrant the most resource allocation for mitigation? This is a good example of how these threat modeling approaches are not contradictory. You can apply multiple models to the same organization. STRIDE helps you identify specific threats and DREAD helps you to identify which threats warrant the most resources.

SQUARE

Security Quality Requirements Engineering is about using requirements engineering from a security perspective. Requirements engineering is a well-established part of systems engineering. SQUARE extends that to security.

In systems engineering requirements engineering is used to define the requirements for the system to be developed. The process is to begin with the informal, and often vague articulation of requirements as per the stakeholders, and to process that into specific and actionable system requirements. In cybersecurity, requirements engineering is a critical component that is often overlooked. Many cybersecurity projects are done simply because they meet minimum requirements for some regulatory requirement or because they are common cybersecurity tasks. SQUARE emphasizes the requirements engineering process.

VAST

VAST Visual, Agile, Simple, Threat Modeling. This threat modeling method is used to enumerate and prioritize threats. VAST is about threat modeling through the software development lifecycle, particularly in Agile programming. VAST works with two concurrent types of models. The application threat model and the operational threat model. Threats are reviewed from both perspectives. Process flow diagrams are used to examine application threats. Data flow diagrams are used to examine operational threats. VAST integrates well with DevOps lifecycles.

PASTA

PASTA: Process for Attack Simulation and Threat Analysis. This is a seven-step methodology for evaluating risk. As the name suggests, it is about simulating attacks in order to analyze the threats. PASTA is risk-centric and it was developed in 2012. The seven stages of PASTA are shown in the following Table 10.1.

PASTA provides a systematic approach to simulating attacks and analyzing threats.

LINDDUN

As with the other models, LINDDUN is an acronym for Link-ability, identifiability, nonrepudiation, detectability, disclosure of information, unawareness, noncompliance. The focus of LINDDUN is on privacy issues. The LINDDUN process starts with a data flow diagram that will examine the systems data flows, data stores, processes, and any external entities.

Attack Trees

Attack trees were mentioned earlier in this chapter. The attack tree describes an attack based on a tree form. The tree root is the goal for the attack. The various limbs of the tree are methods for approaching that root.

Each of these models has proponents and detractors, but the reality is that each has specific strengths scenarios in which it is the appropriate choice. It should also be noted that these approaches are not necessarily mutually exclusive. As an example, it would be absolutely valid to combine the SQUARE emphasis on requirements engineering, with the PASTA risk-centric approach, along with a STRIDE review of threats.

Table 10.1 Pasta

Define objectives	Identify business objectives Identify security & compliance requirements Perform business impact analysis
Define technical scope	Determine the boundaries of the technical environment Capture infrastructure dependencies
Application decomposition	Identify use cases Define entry points and trust levels Identify threat actors Perform data flow diagraming Determine trust boundaries
Threat analysis	Examine probabilistic attack scenarios Perform regression analysis on security events Perform threat intelligence correlation
Vulnerability & weakness analysis	Review existing vulnerability reports Analyze design flaws and abuse cases Review scorings such as CVSS and CVE
Attack modeling	Perform attack surface analysis Perform attack tree development Match vulnerabilities and exploits to attack trees
Risk & impact Analysis	Qualify and quantify business impact analysis Identify countermeasures Perform residual risk analysis Identify risk mitigation strategies

Regardless of the threat model approach, you take, or some hybrid of models, there are some fundamentals that must occur. You always begin by identifying what assets you are trying to protect, what are the threats to those assets, and what specific vulnerabilities might be exploited. Beginning with assets/threats/vulnerabilities at least provides one with an understanding of the situation.

Now that the threats are identified, those threats must be examined in more detail. What are the attack vectors that could be used? What actors would most likely exploit the vulnerabilities? What is the probability of an attack and what is the impact? That final issue, the impact, will be addressed later in this chapter in business impact analysis (BIA).

Terms

There is a specific terminology used in threat modeling, Key terms are defined in this section.

Asset	An asset is a resource of value. It varies by perspective. To your business, an asset might be the availability of information, or the information itself, such as customer data. It might be intangible, such as your company's reputation.
Abuse Case	Deliberate abuse of use case in order to produce unintended results.
Actor	Legit or adverse caller of use or abuse cases.
Attack	An attack is an action taken that utilizes one or more vulnerabilities to realize a threat.
Attack Surface	Logical area exposed to threats and underlying attack patterns.
Attack Tree	Diagram of relationship amongst asset-actor-use case-abuse case-vulnerability-exploit-countermeasure.
Attack Vector	The path of the attack. In other words, how the attack gets into the system.
Audit	A review of system security, often to ensure compliance with some regulatory, legal, or contractual obligations.
Behavioral Indicator	An observable action that provides evidence of an underlying element that may be relevant to a threat assessment and management case.
Black Swan Event	An extremely rare event.
Breach	A successful attack.
Countermeasure	Countermeasures address vulnerabilities to reduce the probability of attacks or the impacts of threats. They do not directly address threats; instead, they address the factors that define the threats.
Defect	A bug or flaw in a system. Often this can be the source of a vulnerability.
Exploit	In the case of threat modeling, often used synonymously with attack.
Impact	Negative value sustained by successful attack(s).
Residual Risk	This is the risk leftover after mitigation has been applied.

Threat	A threat is an undesired event. A potential occurrence, often best described as an effect that might damage or compromise an asset or objective.
Use Case	Functional, as the designed function of an application.
Vulnerability	A vulnerability is a software/firmware code imperfection at the system, network, or framework level that makes an exploit possible.

Tools

In addition to the aforementioned models, it is important to have tools that aid security professionals in being able to acquire threat intelligence. The following sections provide a general overview of specific threat modeling tools.

National Vulnerability Database

In conducting any threat model, it is critical to know what vulnerabilities currently exist. An excellent place to start is the National Vulnerability Database. This is a United States Government repository of vulnerability data using the Security Content Automation Protocol (SCAP). The protocol is usually pronounced as "ess-cap". It is essentially a set of open standards that allows interoperability among vulnerability scanners. The SCAP protocol is used to enable automated vulnerability management. It can also be used to support policy compliance, including compliance with the Federal Information Security Modernization Act of 2002 commonly referred to as FISMA.

In Fig. 10.2, you can see a specific vulnerability in NVD.

This allows one to view a summary of data regarding a particular vulnerability. In this case, the vulnerability is Fig. 10.1: National Vulnerability Database and it is rated as having a medium score. Details regarding impact score, exploitability score, Attack Vector, attack complexity and other properties of the vulnerability are provided. Each vulnerability also as 'additional information'. For this particular vulnerability that is as follows:

- Victim must voluntarily interact with attack mechanism
- Allows unauthorized modification
- Allows disruption of service

Using the National Vulnerability Database is an excellent first step in performing threat assessments. It is also important to ensure that any vulnerability management or scanning tool you utilize is compliant with the SCAP protocol, to ensure interoperability.

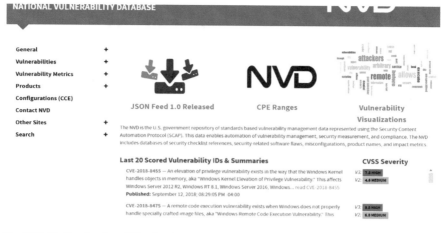

Fig. 10.1 National vulnerability database

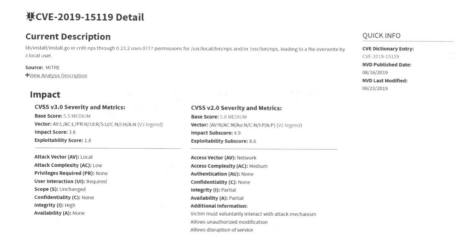

Fig. 10.2 Specific vulnerability

US CERT

https://www.us-cert.gov/ncas/alerts

CERT is managed by the US Department of Homeland Security. The purpose of CERT is primarily information exchange, but they also engage in training, vulnerability assessments, data synthesis, and related activities. They define a computer security incident as "A computer security incident within the US Federal Government is defined by CISA and the US National Institute of Standards and Technology

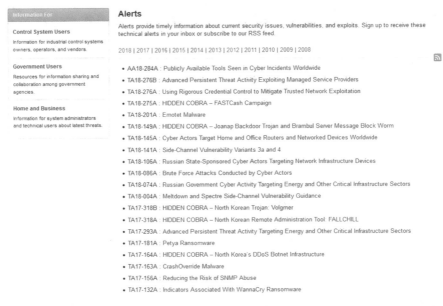

Information For

Control System Users
Information for industrial control systems owners, operators, and vendors.

Government Users
Resources for information sharing and collaboration among government agencies.

Home and Business
Information for system administrators and technical users about latest threats.

Alerts

Alerts provide timely information about current security issues, vulnerabilities, and exploits. Sign up to receive these technical alerts in your inbox or subscribe to our RSS feed.

2018 | 2017 | 2016 | 2015 | 2014 | 2013 | 2012 | 2011 | 2010 | 2009 | 2008

- AA18-284A : Publicly Available Tools Seen in Cyber Incidents Worldwide
- TA18-276B : Advanced Persistent Threat Activity Exploiting Managed Service Providers
- TA18-276A : Using Rigorous Credential Control to Mitigate Trusted Network Exploitation
- TA18-275A : HIDDEN COBRA – FASTCash Campaign
- TA18-201A : Emotet Malware
- TA18-149A : HIDDEN COBRA – Joanap Backdoor Trojan and Brambul Server Message Block Worm
- TA18-145A : Cyber Actors Target Home and Office Routers and Networked Devices Worldwide
- TA18-141A : Side-Channel Vulnerability Variants 3a and 4
- TA18-106A : Russian State-Sponsored Cyber Actors Targeting Network Infrastructure Devices
- TA18-086A : Brute Force Attacks Conducted by Cyber Actors
- TA18-074A : Russian Government Cyber Activity Targeting Energy and Other Critical Infrastructure Sectors
- TA18-004A : Meltdown and Spectre Side-Channel Vulnerability Guidance
- TA17-318B : HIDDEN COBRA – North Korean Trojan: Volgmer
- TA17-318A : HIDDEN COBRA – North Korean Remote Administration Tool: FALLCHILL
- TA17-293A : Advanced Persistent Threat Activity Targeting Energy and Other Critical Infrastructure Sectors
- TA17-181A : Petya Ransomware
- TA17-164A : HIDDEN COBRA – North Korea's DDoS Botnet Infrastructure
- TA17-163A : CrashOverride Malware
- TA17-156A : Reducing the Risk of SNMP Abuse
- TA17-132A : Indicators Associated With WannaCry Ransomware

Fig. 10.3   CERT

as a violation or imminent threat of violation of computer security policies, acceptable use policies, or standard security practices."[1] An example of CERT alerts is shown in Fig. 10.3.

SHODAN

https://www.shodan.io
SHODAN is a tool that is widely used in the cybersecurity community. Shodan is essentially a search engine that allows one to search for any vulnerable system that is connected to the internet. Shodan collects data on devices and services connected to the internet and provides an interface for security personnel to search it. This project was launched in 2009 by John Matherly (Fig. 10.4).

There are many options you can use in searching with Shodan.io, some are given here:

Search for default passwords
default password country:US
default password hostname:chuckeasttom.com
default password city:Chicago
Find Apache servers
apache city:"San Francisco"

---

[1] https://www.us-cert.gov/about-us

Fig. 10.4 Shodan

Find Webcams
webcamxp city:Chicago
OLD IIS
"iis/5.0"
The preceding list are examples of search terms, the filters you can use include

city   find devices in a specific city
country   find devices in a specific country
geo   you can pass it coordinates (i.e., latitude and longitude)
hostname   find values that match a specific hostname
net   search based on an IP or/x CIDR
os   search based on operating system
port   find particular ports that are open
before/after   find results within a timeframe

Shodan is really quite easy to use. It is an excellent place to begin a threat assessment. Particularly as Shodan allows one to limit searches to a particular domain. Thus, searching the domain in question to determine if any vulnerabilities are shown in Shodan is an easy first step.

Threat Crowd

https://www.threatcrowd.org
This search site is less robust than Shodan but still useful. One can search any domain, IP address, email address, or an organization, this lets you know if that particular search target is associated with any known threats. The landing page is shown in Fig. 10.5.

Fig. 10.5 Threatcrowd

One common way this is used is with suspicious emails. The email domain name, and IP addresses from the header can be searched to determine if they are associated with threats. This same process can be done for suspect websites. This is more of a tactical threat analysis tool in that it let you check specific and immediate threats.

Common Vulnerability Scoring System

The common vulnerability scoring system (CVSS) is widely used to classify vulnerabilities. This is an open industry standard that allows for the scoring of vulnerabilities based on severity. The full specification can be found here https://www.first.org/cvss/specification-document.

There are three groups of metrics: base, temporal, and environmental. The base group describes the basic characteristics of the vulnerability that are not determined by time (temporal) or environment. The metrics in this group are Attack Vector, Attack Complexity, Privileges Required, User Interaction, Scope, Confidentiality Impact, Integrity Impact, and Availability Impact.

The Attack Vector Metric can be Network (N), Adjacent (A), Local (L), Physical (P). Attack Complexity can be None (N), Low (L), and High (H). User Interaction can be: None (N) or Required (R). The Scope metric captures whether a vulnerability in one vulnerable component impacts resources in components beyond its security scope. Its values can be: Unchanged (U) or Changed (C). The Impact Metrics (Confidentiality, Availability, or Integrity) are all rated: High (H), Low (L), or None (N).

The Temporal Metric Group has three metrics: Exploit Code Maturity, Remediation Level, and Report Confidence. The Environmental Metric Group has four metrics: Modified Base Metrics, Confidentiality Requirement, Integrity Requirement, and Availability Requirement.

Exploit Code Maturity measures the likelihood of the vulnerability being attacked and is typically based on the current state of exploit techniques, exploit code availability, or active, "in-the-wild" exploitation. The possible ratings are: Not Defined (X), High (H), Functional (F), Proof of Concept (P), and Unproven (U). The Remediation Level Metric can be: Not Defined (X), Unavailable (U), Workaround (W),

Temporary Fix (T) or Official Fix (O). The Report Confidence metric indicates how confident we are in the details of the vulnerability. Its values can be: Not Defined (X), Confirmed (C), Reasonable (R), and Unknown (U).

The values for the various metrics are summarized in the following Table 10.2:

CVSS   scoring   is   often   represented   as   a   string   such   as `CVSS:3.1/S:U/AV:N/AC:L/PR:H/UI:N/C:L/I:L/A:N/E:F/RL:X`

As can be seen, CVSS scoring is a bit more involved than some of the other methods, such as CVE. However, it is also more informative. This method provides a quantifiable approach to categorizing vulnerabilities so they can be appropriately addressed. It is strongly recommended that you become well acquainted with CVSS.

Table 10.2 CVSS metrics

Metric group	Metric name (and abbreviated form)	Possible values	Mandatory?
Base	Attack Vector (AV)	[N,A,L,P]	Yes
	Attack Complexity (AC)	[L,H]	Yes
	Privileges Required (PR)	[N,L,H]	Yes
	User Interaction (UI)	[N,R]	Yes
	Scope (S)	[U,C]	Yes
	Confidentiality (C)	[H,L,N]	Yes
	Integrity (I)	[H,L,N]	Yes
	Availability (A)	[H,L,N]	Yes
Temporal	Exploit Code Maturity (E)	[X,H,F,P,U]	No
	Remediation Level (RL)	[X,U,W,T,O]	No
	Report Confidence (RC)	[X,C,R,U]	No
Environmental	Confidentiality Requirement (CR)	[X,H,M,L]	No
	Integrity Requirement (IR)	[X,H,M,L]	No
	Availability Requirement (AR)	[X,H,M,L]	No
	Modified Attack Vector (MAV)	[X,N,A,L,P]	No
	Modified Attack Complexity (MAC)	[X,L,H]	No
	Modified Privileges Required (MPR)	[X,N,L,H]	No
	Modified User Interaction (MUI)	[X,N,R]	No
	Modified Scope (MS)	[X,U,C]	No
	Modified Confidentiality (MC)	[X,N,L,H]	No
	Modified Integrity (MI)	[X,N,L,H]	No
	Modified Availability (MA)	[X,N,L,H]	No

OSSTMM

The Open Source Security Testing Methodology Manual (OSSTMM) is a manual for testing security. It is, as the name suggests open source. The standard can be found at http://www.isecom.org/research/ OSSTMM uses the concept of modules, defining them as a set of processes or phases which are applicable for each channel. The four modules/phases are

- Phase I: Regulatory
- Phase II: Definitions
- Phase III: Information
- Phase IV: Interactive controls test

OSSTMM focuses on which items need to be tested, what to do before, during, and after a security test, and how to measure the results

Business Impact Analysis

Several threat models discussed earlier in this chapter did explicitly mention business impact analysis. However, even those models that do not explicitly use BIA, can incorporate a BIA. The concept is to analyzed what impact a given realized threat would have on the organization. Consider a web server crash. If your organization is an e-commerce business, then a web server crash is a very serious disaster. However, if your business is a restaurant and the website is just a way for new customers to find you, then a web server crash has a lesser impact. You can still do business and earn revenue while the web server is down. You should make a spreadsheet of various likely or plausible disasters and do a basic BIA for each.

A few things go into your BIA. There are explicit quantifiable metrics that go into the BIA. One item to consider is the maximum tolerable downtime (MTD). How long can a given system be down before the effect is catastrophic and the business is unlikely to recover? This should be as objective as possible, and of course, you will want to err on the side of caution.

Another item to consider in the BIA is the mean time to repair (MTTR). This is sometimes also called mean time to recover. How long is it likely to take to repair a given system if it is down? With computer hardware and equipment, you will often be able to quantify an average (i.e., mean) time required to repair a given device.

Another metric that is part of BIA is mean time before failure (MTBF). This is also sometimes called mean time between failures. It is an estimate of how long a given system will run before a failure is likely. With computer hardware, manufacturers often have statistics on failure rates. These are obviously means. Some individual devices might have a lot longer lifespan before failure, others might fail sooner.

Consider a file server. If one determines that the MTTR is 48 h, then one has to consider the impact on the business for that file server being offline for 48 h. This metric is further influenced by considering the MTBF. If the MTBF is 10 years, then the file server being offline is less of a concern than if the MTBF is 2 years. This is a rather simplistic example, but it illustrates the role that these metrics have on a business impact analysis.

There are other metrics that are helpful in business impact analysis. When considering a specific threat, you should also consider the Single Loss Expectancy (SLE) and Annualized Loss Expectance (ALE). The Single Loss Expectancy (SLE) is the cost of a single loss. SLE is the Asset Value (AV) times the Exposure Factor (EF). Put as a formula that is:

SLE = AV * EF

Now, this brings us to the question of what is the exposure factor. This is an estimate of how much will be lost. Consider a webserver used for e-commerce. If the web server is offline for 10% of a given time period, then it has lost 10% of its worth. Its exposure factor is 10%.

The Annualized Loss Expectancy (ALE) is your yearly cost due to risk. It is calculated by multiplying the Single Loss Expectancy (SLE) times the Annual Rate of Occurrence (ARO). Put as a formula this is:

ALE = SLE & ARO

To illustrate this, let us return to the example of a webserver used for e-commerce. Assume this server earns 30,000 US dollars per month. Further, assume that a denial of service attack will keep the computer offline for approximately 1 full day before service can be restored. Using 30 days as an average month length this means the exposure factor due to a DoS attack would be 1/30 or 3.33%. So, the exposure factor (EF) is 3.33%. That means that a single loss would be 30,000 * 0.0333 or 1,000 US dollars. To compute the annualized loss expectance (ALE) look at statistics for industry, region, past incident response, etc. to determine what the likely Annualized Rate of Occurrence (ARO) is. To calculate ALE the SLE * ARO is the formula. So, if the analysis determines ARO is 3, then the ALE would be 1000 (SLE) * 3 (ARO) or $3000.

This particular scenario is rather simplified and only accounts for a single server. It also does not account for issues such as loss of reputation. However, this does provide a quantifiable piece of data to incorporate in the business impact analysis (BIA).

Characterizing and Analyzing Network Traffic

Being able to understand network traffic is a fundamental part of threat assessment. Often network attacks are detected by Intrusion Detection Systems (IDS), but a security professional still needs to understand how to analyze the network traffic themselves.

RMON

The Remote Network MONitoring (RMON) was developed by the Internet Engineering Task Force (IETF) in order to support network monitoring and protocol analysis. RMON is a standard monitoring specification that allows various network monitors to exchange network monitoring data.

The original version of RMON had 10 groups:

1. Statistics: real-time LAN statistics, e.g., utilization, collisions, CRC errors
2. History: history of selected statistics
3. Alarm: definitions for RMON SNMP traps to be sent when statistics exceed defined thresholds
4. Hosts: host-specific LAN statistics, e.g., bytes sent/received, frames sent/received
5. Hosts top N: record of N most active connections over a given time period
6. Matrix: the sent-received traffic matrix between systems
7. Filter: defines packet data patterns of interest, e.g., MAC address or TCP port
8. Capture: collect and forward packets matching the Filter
9. Event: send alerts (SNMP traps) for the Alarm group
10. Token Ring: extensions specific to Token Ring.

That was then expanded to 10 more:

1. Statistics: real-time LAN statistics, e.g., utilization, collisions, CRC errors
2. History: history of selected statistics
3. Alarm: definitions for RMON SNMP traps to be sent when statistics exceed defined thresholds
4. Hosts: host-specific LAN statistics, e.g., bytes sent/received, frames sent/received
5. Hosts top N: record of N most active connections over a given time period
6. Matrix: the sent-received traffic matrix between systems
7. Filter: defines packet data patterns of interest, e.g., MAC address or TCP port
8. Capture: collect and forward packets matching the Filter
9. Event: send alerts (SNMP traps) for the Alarm group
10. Token Ring: extensions specific to Token Ring

The original RMON was defined by RFC 2819. RMON2 was defined in RFC 4502. There have been modifications of RMON for specialized networks. For example, Remote Network Monitoring Management Information Base for High Capacity Networks or HCRMON was defined in RFC 3272. SIMON or Remote Network Monitoring MIB Extensions for Switched Network was defined in RFC 2613.

Wireshark

Wireshark provides a convenient graphical user interface (GUI) for examining network traffic. It is a free download you can get at https://www.wireshark.org/. You can see the main interface in Fig. 10.6.

This easy to use graphical interface is one reason Wireshark is so popular. It is also a free product, thus fits into any budget. It is easy to highlight packets and analyze

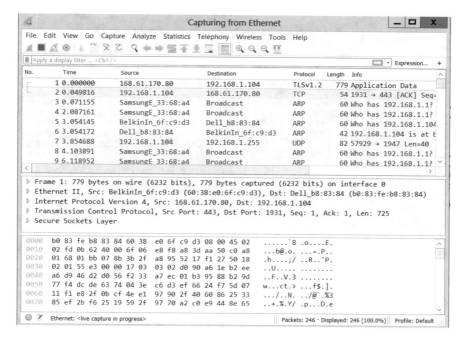

Fig. 10.6 Wireshark main interface

network traffic using Wireshark. You can highlight any individual packet and analyze it. That is shown in Fig. 10.7.

When using Wireshark, you will quickly see that there is a very large amount of data collected. One way to deal with the copious amount of data, is to filter what you are viewing. Display filters (also called post-filters) filter the view of what you are seeing but don't actually eliminate any packets from the capture. All packets in

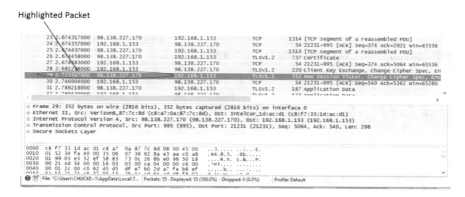

Fig. 10.7 Individual packets

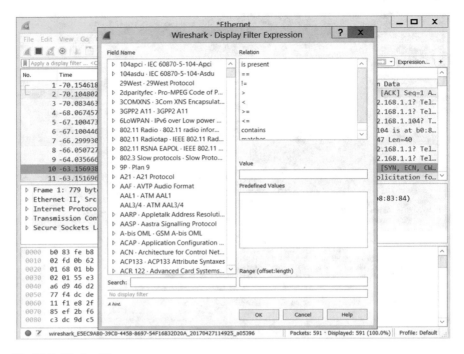

Fig. 10.8   Display filters

the capture still exist in the trace. Display filters use their own format and are much more powerful then capture filters. A display filter is shown in Fig. 10.8.

There are a number of advanced tools available from the Wireshark dropdown menus. One is the ability to follow a particular TCP steam. You can highlight a single packet the follow the entire TCP stream related to that packet. This is shown in Fig. 10.9.

This is just an introduction to Wireshark. Entire books could be written (and have been) on Wireshark. The purpose of this section is to introduce you to this tool. Wireshark has an online guide https://www.wireshark.org/download/docs/user-guide.pdf and there are numerous tutorials around the internet.

Cisco Log Levels

Cisco devices are ubiquitous, particularly in the United States and Europe. These devices have a defined log level that is used to indicate issues in the log. Since so many switches, routers, and other devices are Cisco, or at least compliant with Cisco, it is important to understand the Cisco logs in order to understand threats. The log levels are shown in Table 10.3.

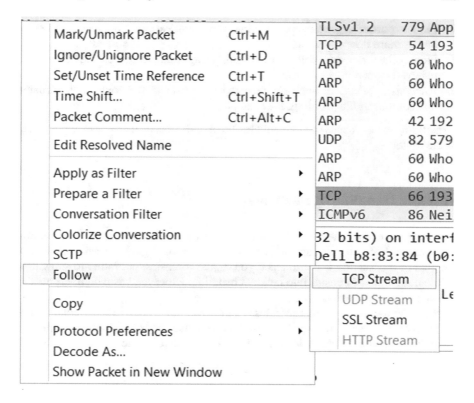

Fig. 10.9 Follow TCP stream

Table 10.3 Cisco log levels

Level	Description
0—emergency	System unusable
1—alert	Immediate action needed
2—critical	Critical condition
3—error	Error condition
4—warning	Warning condition
5—notification	Normal but significant condition
6—informational	Informational message only
7—debugging	Appears during debugging only

NetFlow

NetFlow is another creation of Cisco. It was introduced on Cisco routers in 1996 and it provides the ability to collect IP network traffic as it enters or exits the interface. There are three primary components:

1. Flow exporter: aggregates packets into flows and exports flow records towards one or more flow collectors.
2. Flow collector: responsible for reception, storage and pre-processing of flow data received from a flow exporter.
3. Analysis application: analyzes received flow data in the context of intrusion detection or traffic profiling, for example.

NetFlow is one means of analyzing traffic in real-time, as it crosses the interfaces of the router.

Forensic Handling of Incidents

Clearly a complete coverage of digital forensics is beyond the scope of a section in a chapter. Entire books can and have been written on digital forensics. However, the goal of this section is not to teach you digital forensics, but rather to show how to integrate digital forensics into incident handling. However, basic digital forensics principles will be covered.

When an incident occurs, regardless of the level or severity of the incident, their needs to be an organized response. And that response needs to include not only the immediate reaction to the incident itself but a forensic response.

Defining Forensics

Digital forensics is a scientific discipline that seeks to gather evidence that can be utilized in court. Now when doing incident response, it is frequently the case that you are not thinking about the court, either civil or criminal. However, that is always a possible outcome. You might be investigating a malware incident, and your primary focus is to restore normal operations. Then in the course of the investigation, you discover the malware was placed on the system by a disgruntled employee. Now you wish to pursue criminal charges and/or civil litigation. If you have mishandled the evidence, then that won't be possible, or will at least be severely hampered.

Digital forensics does utilize scientific techniques, but in a manner that is consistent with legal requirements. This begins with a chain of custody. You must document the disposition of evidence from the moment evidence is discovered until it is presented in a court. You must document the seizure of the evidence and everyone who had access to it. It is also important to secure the evidence. At a bare minimum, this would be placing it in a locked cabinet to which few people have access. We will address specific forensic issues as we walk through the four steps of incident response.

Four Steps

Incident response can certainly be a complex process. How to stop the incident alone can be cumbersome. However, all incident response can be distilled into four phases. Those phases are containment, eradication, recovery, and follow-up. In this section, we will examine each of these along with the appropriate forensic steps to take.

Containment

The first step is always to limit the incident, to prevent additional systems from being affected. Consider the example of a virus. The first step is to disconnect the infected device so the virus does not spread further. Other incidents may not have such a clear containment path. For example, if there is an intruder getting into the web server how is that contained? Well, the first step would be to isolate that server.

During the containment phase, it is a good time to image a device. Forensic imaging involves making a bit by bit copy of the device's storage. In the case of malware, there may be a large number of machines infected. You do not need, nor really want to image them all. Instead image at least one. There are many tools that can image a drive. Most major forensics tools will image the drive.

Let us consider one particular forensic tool. This tool is chosen because it has a free trial download that you can use and try this yourself. The tool is OSForensics, the website for this tool is https://www.osforensics.com/. Imaging is simple. First select forensic imaging, as shown in Fig. 10.10.

Now you simply select where to copy the image to. Usually, this will be to an external hard drive. Always select the hash function. That is used to verify that the

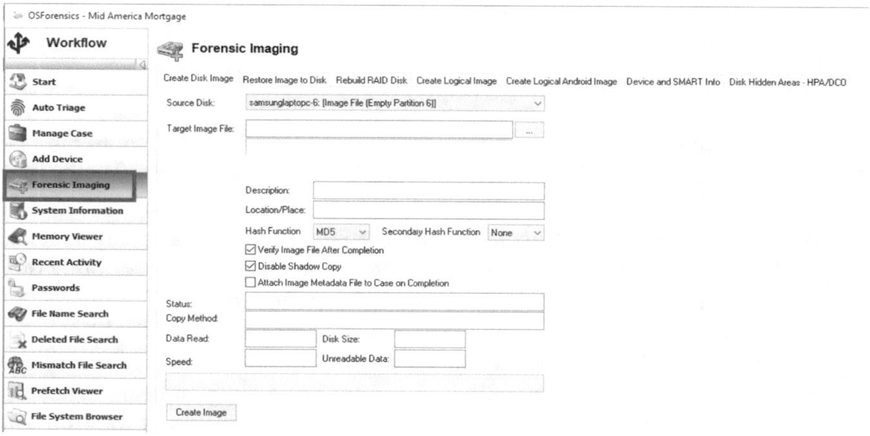

Fig. 10.10 Imaging with OSForensics

Fig. 10.11 Install OSForensics to A USB

copy is valid. The original drive and the image are hashed and the hashes compared. This provides evidence that nothing went wrong in copying. To image a machine, you will usually install OSForensics to a USB drive, and plug that USB drive into the suspect machine. That is easy to do and can be seen in Fig. 10.11

Once you have imaged the drive, you can secure that image in some locked storage. Now you can focus on just incident response, knowing you have an image of the suspect machine that you can forensically examine later. Failure to image the machine at this early stage may lead to you being unable to get a forensic image later.

Eradication

Once the incident is contained, the next step is to eradicate the problem. In the case of malware, the issue is to remove the malware. In many cases, anti-malware such as Norton, Bit Defender, McAfee, Kaspersky or AVG can remove the malware. In other situations, it will be necessary to manually remove the malware. In any incident, the goal of eradication is to remove the problem. If that is cross-site scripting on a company website, then eradication involves removing the script.

If you did not image the machine in the containment phase, it must be the first thing you do in this phase. If the vulnerability is simply eradicated, it is likely that evidence will be eradicated along with it. It is imperative that you begin collecting evidence prior to eradicating the vulnerability. Again, the analysis of forensic data need not be done at this stage. But you must capture forensic data. In addition to imaging drives, this is a good time to make backup copies of relevant logs.

Recovery

Recovery involves returning the affected systems to normal status. In the case of malware that means ensuring the system is back in full working order with absolutely no presence of the malware. In many cases, this will involve restoring software and data from a backup source that has been verified to be free from the malware infection. This stage is not important forensically. You might start the forensic analysis in parallel with this stage. But forensics and recovery are two separate processes.

Follow-up

The follow-up phase is another stage at which forensics plays a critical role. If you had a breach that will lead to criminal or civil litigation, then this is where you will be gathering all that evidence. But even if you are not moving towards litigation, this is the phase at which you determine the full cause of the event.

Regardless of the specifics of the incident, it is critical that the evidence is preserved. The usual emphasis for corporate disaster recovery is simply a return to normal operations as soon as possible. Frequently this is done at the expense of preserving forensic evidence. This can lead to many problems. First and foremost, failure to preserve forensic information will prevent the response team from effectively evaluating the cause of the incident and adjusting company policies and procedures to reduce the risk of such an incident being repeated. Even if the incident does not involve a crime or the company simply does not wish to prosecute, forensic data is an integral part in preventing future incidents.

Forensic Preparation

Realizing the importance of forensics in incident response is an important first step. But this realization still leaves the question of how to implement proper forensics procedures. There are specific steps that an IT department can take to intertwine forensic techniques with the company's incident response policies.

Forensic Resources

The first step is to identify forensic resources that the organization can utilize in case of an incident. No amount of policy changes will be effective if the company does not have access to forensically trained individuals. One approach an organization can take is to get basic forensics training for its own IT security staff. Many college computer-related degrees now include forensics courses and most security-related degrees include at least an introductory forensics course. If no one on the company's IT security staff has had such training it may be helpful to send staff members to be trained in computer forensics and perhaps to obtain one of the major forensics' certifications.

Another option the organization can pursue is to identify an outside party that can respond to incidents with forensically trained personnel. In this case, part of incident planning would involve ensuring there is an agreement in place with a reliable forensics company or an individual consultant. If this is the option an organization wishes to pursue it is critical to ensure that the organization identified has both an appropriate level of competency and has the resources to respond to incidents.

Forensics and Policy

Once appropriate forensic resources have been identified, forensic methodology must be integrated into the incident response policy for the organization. This means that all policies regarding incident response will need to be updated. Who will seize evidence? What tools will be utilized? Where will evidence be stored? Who will be briefed on the forensic investigation? These are all questions that need to be clearly addressed in policies. Even if the procedure is to call an outside forensic consultant, the staff needs to know who to call and when.

It is likely that even if the IT security staff are not trained specifically in forensics, they have some basic knowledge of the field. The reason is that many security textbooks now include at least a chapter on basic forensics. Most of the general computer security certifications such as CompTIA Security + and CISSP also include sections on basic forensics. Even if your staff lacks the appropriate training to perform a forensic investigation, they should be trained well enough to know how to preserve evidence and avoid any alteration of the evidence.

Conclusions

This chapter focused primarily on risk handling. Modeling risks and ranking them was addressed first. You were also introduced to cyber-threat intelligence. A wide range of tools, techniques, and thread models were introduced. Finally, we discussed the essentials of digital forensics, and how to integrate that into your incident response policy.

Chapter 11
Training, Education, and Awareness

Izzat Alsmadi

K0208: Knowledge of Computer-Based Training and e-Learning Services

Computer-Based Training (CBT) involves the use of computers or computing environments for delivery and access to training courses or programs. CBT can take many different formats:

- Synchronous, where educators and students meet, live regardless of their physical location. This is probably the closest environment to classical face to face (F2F) teaching or training.
- Asynchronous, not live: Educators create the training on their own time, and students listen to the educational content on their own time.
- Hybrid, between synchronous, asynchronous or between online and F2F. Online coursework may be offered as hybrid learning or blended learning experiences that are offered partially online and partially in the formal classroom.
- Online, web-based, mobile, and distance learning. Different terminologies that are all used to refer to the same context with slight variations related to the environment that educators or students are using to create or access the educational content.

There are many advantages for both educators and learning on CBT in comparison with classical training. The main advantage is convenience and flexibility: No doubt that the significant advantage of going CBT as an alternative to F2F is that it is more convenient for both educators and students. Educators can teach from any location in the world and so as students. They can bypass all physical, border control, etc. barriers and be able to teach or learn from anywhere in the world. This means also saving on travel costs, relocation, etc. CBT opens new opportunities for many institutions to offer courses and programs for a broad spectrum of candidate students beyond the local students in their city or state. Universities can also hire remote faculties which can save directly or indirectly in the budget.

I. Alsmadi et al., *The NICE Cyber Security Framework*,
https://doi.org/10.1007/978-3-030-41987-5_11

Learners can take the course content based on their preferences and time. With F2F environments, this can be more challenging. For example, a student can rewind a recorded lecture again and again or repeat individual sections for further focus or understanding. If it's not live, they may not have the ability to clarify from their instructors' unclear content. Nonetheless, most CBT environments enable different methods of communication between educators and learners for feedback, questions, etc.

The advances of the Internet (e.g., getting larger and larger bandwidths) and the advances of media tools and applications facilitate making CBT easier and more reliable.

CBT can also make education standardized and consistent. Streaming and media applications are now integrated with online learning environments (e.g., BlackBoard, Moodle, etc.).

The relation between educators and learners is more flexible in CBT environments. Students these days not only take courses from their primary institution or University, but they can also take courses from the online course offering vendors (e.g., Udemy, Udacity, Coursera, Alison, Creative Live, EdX, Canvas Network, eduCBA, etc.). Individual users can also create educational channels on websites such as YouTube and create educational content in any field or subject.

Opponents of CBT cite issues related to the quality of education, plagiarism, security issues, integrity issues, etc. However, technology trends and reality are shifting training gradually towards such directions. Learning institutions will have to deal with such reality and find techniques to deal with their problems or limitations.

K0215: Knowledge of Organizational Training Policies

Organizations understand that training helps them remain competitive by continually educating their employees. They understand that investing in their employees yields higher productivity. However, training is not as intuitive as it may seem. Studies indicate that there are right and wrong ways to design, deliver, and implement a training program.

Organizational training, policies, and procedures can ensure that all employees know how to behave and respond appropriately to any situation. Training structures and practices can support the development of learning flexible manners, developing technologies for learning, and making learning an articulated priority within the context of production goals and targets. Smart decisions must be made about where and how funds are allocated and which policies should be supported to help in workforce development and training. Policymakers and managers should understand how practical workforce training and development can ensure a knowledgeable and skilled workforce. Resources are not limitless, so policies should help encourage wise investments, and public funding should target focused training programs that are likely to be successful.

In addition to (1) the support from managers and human resources and policies, workers also need (2) proper and relevant resources, including access to information (i.e., manuals, coaches, access to ongoing sources of information in the field; (3) abilities to visits to other relevant training locations and worksites; (4) workshops, conferences and short courses; and (5) support for independent study to earn qualifications or certifications.

Organizations should support their employees to earn relevant certificates to their work roles. Industry certifications and licenses are credentials obtained through an examination process. Industry certifications and licenses allow learners to document and verify their skills and knowledge through a formal examination process. Certifications are frequently offered as part of a traditional educational program but can be obtained by individuals regardless of whether they have taken a traditional educational program or developed their skills and knowledge.

Work-based learning or training is learning that occurs on the job with skills and knowledge related to the company and the nature of the job. It provides opportunities for socialization into a profession and is useful for clarifying career choices. Several challenges exist in the quality of work-based learning include that training is not standardized and may not be aligned with the curriculum being taught.

Most middle to large companies have their own training departments in which they create and deliver training content to their employees. The training targets not only domain-specific material related to the company business functions but also general training related to company policies and regulations, human resource issues, general information security subjects, and so on. Some of the training materials are generic to apply to and are required by all company employees. Other training materials target individual employees with specific roles or business functions.

Different studies indicated that organizational training activities are recognized as being useful in providing a competitive advantage through their impact on the employees' productivity, which is achieved by improving employees' skills and performance and by inducing positive behavioral changes. Employees must be motivated by specific means to pursue continuous skills' development activities. They can be motivated by using different options. Rewards can be in the form of career advancement. While premised on the notion that a worker can advance his or her career within an organization, this strategy is becoming less common. Beyond career advancement, this strategy has also been linked to better wages and increased productivity in the workplace. Companies should establish appropriate attendance policies while recognizing the need to balance between mandatory and optional training requirements can be tricky.

Learn and Earn Models: Some companies provide opportunities for their employees to earn higher education degrees while maintaining their careers. Expenses of their education in most cases are covered through their work. These programs take forms ranging from internship and co-ops to work-study programs, career, and technical education.

Training programs should be continually monitored and evaluated. Training should be evaluated to determine if it was successful and met training objectives.

Feedback should be obtained from learners to determine program and instructor effectiveness and also knowledge or skill acquisition.

K0216: Knowledge of Learning Levels (i.e., Bloom's Taxonomy of Learning)

Bloom's taxonomy can be used to: create learning assessments, plan lessons, evaluate the complexity of assignments, design curriculum contents, develop courses, etc. The belief is that learners move up through each level of the pyramid in Bloom's taxonomy, starting from fundamental learning, to acquiring more in-depth knowledge on a subject, with each level crucial to the development of the next level.

Since it was proposed in 1956, Bloom taxonomy became a popular model to show and understand learning levels [1]. A group of cognitive psychologists, curriculum theorists and instructional researchers, and testing and assessment specialists published in 2009 a revision of Bloom's Taxonomy [2, 3] (Fig. 11.1).

Both original and revised Bloom taxonomies identified six learning levels. However, the differences are in the sequence and names of those six different levels. Revised Bloom shifted terminology from nouns to verbs.

The six levels in the revised Bloom are:

1. Remembering: Recognizing or recalling knowledge from memory. This is the lowest learning level in the taxonomy. Remembering is when memory is used to produce or retrieve definitions, facts, or lists, or to recite previously learned information. Direct questions (e.g., multiple-choice, simple essay questions, etc.) are used to evaluate this learning level.

   This level includes sub-levels: Terminology, Specific facts, Conventions, Trends and sequences, classifications and categories, Criteria, Methodology, Principles and generalizations, Theories, and structures.

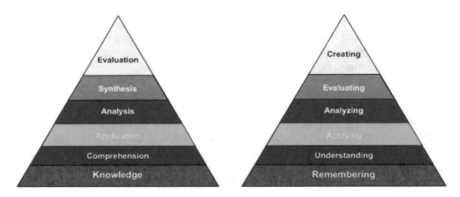

Fig. 11.1 Original (left) versus revised (right) Bloom Taxonomy

2. Understanding: Constructing meaning knowledge from different types of functions, whether they are written or graphic messages. Activities at this level include interpreting, exemplifying, classifying, summarizing, inferring, comparing, or explaining.
3. Applying: Carrying out or using a procedure, project, etc. through executing, or implementing. Applying relates to or refers to situations where learned material is used through products such as models, presentations, interviews or simulations.
4. Analyzing: Breaking materials or concepts into parts, determining how the parts are related to one another or how they interrelate, or how the parts relate to an overall structure or purpose.
5. Evaluating: Making judgments based on criteria and standards through checking and critiquing. Critiques, recommendations, and reports are some of the products that can be created to demonstrate the learning stage of evaluation.
6. Creating: Putting elements together to form a coherent or functional whole. It also includes reorganizing elements into a new pattern or structure through generating, planning, or producing.

The understanding of Bloom taxonomy and learning stages can help us create solutions for problems in the proper sequence:

- Before you can (2) **understand** a concept, you must (1) **remember** it.
- To (3) **apply** a concept, you must first (3) **understand** it.
- To (5) **evaluate** a process, you must have (4) **analyzed** it.
- To (6) **create** an accurate conclusion, you must have completed a thorough (5) **evaluation**.

Critics of Bloom taxonomy have questioned whether human cognition can be divided into distinct categories, sequential or hierarchical categories. Others recognize that it does not—and cannot—represent human thought or learning in all complexity and sophistication.

K0217: Knowledge of Learning Management Systems and Their Use in Managing Learning

Learning Management Systems (LMSs) continue to get much attention, as more colleges and universities launch online programs and design distance learning options for students.

LMS is a web-based software application that delivers, organizes, and manages educational or training content. Students can access course-works or training materials, take quizzes, get real-time feedback, and evaluate courses. Instructors and administrators can build and launch courses, monitor student progress, and generate reports. Figure 11.2 shows the current LMS market share [4].

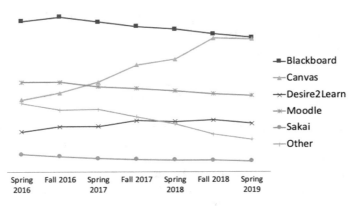

Fig. 11.2   LMS market share [4]

Popular Commercial LMSs

1. Blackboard. WebCT (Course Tools) or Blackboard Learning System, is an online proprietary virtual learning environment system which has the largest LMS market share
2. Desire2Learn
3. Pearson eCollege
4. Canvas.

Open Source LMSs

1. Moodle
2. Sakai
3. Segue
4. Coursework
5. ATutor
6. Claroline.

   Learning management systems are also used by government organizations and private companies. The LMS is a one-stop system for managing learning, compliance, online courses, and policies at your firm and across multiple jurisdictions. Employee training forms a crucial part of quality control as well as governance, risk, and compliance programs for companies, big or small. McIntosh [5] listed 767 active corporate LMSs. Some of the popular corporate LMSs or learning platforms in the industry include:

1. Adobe Captivate Prime
2. LearnUpon LMS
3. Docebo LMS
4. SAP Litmos LMS
5. Looop
6. TalentLMS
7. iSpring Learn
8. Kallidus Learn
9. 360Learning Engagement Platform
10. Northpass.

K0218: Knowledge of Learning Styles (e.g., Assimilator, Auditory, Kinesthetic)

A learning style is a way a person prefers to learn. Learners have better abilities to digest knowledge once they are given their preference or style in learning.

Kolb [6] developed an experiential learning style theory that is comprised of four stages:

1. Getting involved in actual experiences.
2. Reflective observation of the new experience.
3. Developing a new idea with an abstract conceptualization based on reflection.
4. Active experimentation with the new idea.

The four resulting learning styles from Kolb are **Diverger, Assimilator, Converger, and Accommodator**, Fig. 11.3.

McCarthy [7] developed the 4MAT model, identifying four different types of learners:

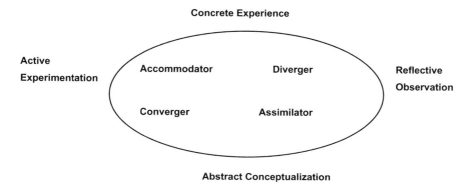

Fig. 11.3 Kolb learning styles [6]

Table 11.1 Barbe and
Milone [9], VAK model

Visual	Kinesthetic/tactile	Auditory
Picture	Gestures	Listening
Shape	Body movements	Rhythms
Sculpture	Object manipulation	Tone
Paintings	Positioning	Chants

1. Learner performs creative learning with a focus on making connections.
2. Learners use analytic learning, focusing on formulating ideas.
3. Learners utilize common sense learning and focus on applying ideas.
4. Learners use dynamic learning, with a focus on creating original adaptations and learning by trial and error.

Gardner [8] defined eight intelligence areas:

1. Linguistic intelligence
2. Logical-mathematical intelligence
3. Musical intelligence
4. Bodily-kinesthetic intelligence
5. Spatial intelligence
6. Interpersonal intelligence
7. Intrapersonal intelligence
8. Naturalistic intelligence.

Barbe and Milone in 1981 proposed three learning modalities, VAK, Table 11.1:

1. Visualizing modality
2. Auditory modality
3. Kinesthetic modality.

K0220: Knowledge of Modes of Learning (e.g., Rote Learning, Observation)

There have been various attempts to classify learning styles, including analytical, concrete, communicative, and authority oriented or visual, auditory, physical, etc.

Modes of learning are a set of roles or guidelines that describe the methods people use to acquire, process, and maintain knowledge. Humans differ in how they learn most effectively; most people favor different combinations of visual, auditory, reading, or kinesthetic (VARK) learning modes.

Rote learning refers to mechanical memorizing without understanding. The notion of rote learning seems to have different meanings in different learning cultures. If the exam heavily depends on memorization, rote learning is the most rational approach to preparing for the examinations.

Learners who use rote learning never really understand or think about what they are learning, which could influence, negatively, any attempts to apply learned information, to new, unfamiliar situations. For the rote learner, new and existing knowledge remains unrelated and meaningless information, which has been memorized, but may quickly be forgotten.

On the other hand, educators who want to test beyond rote learning should design assessments that encourage conceptual understanding as opposed to rote learning. This might be achieved through the increased use of problem-solving, case studies, and the like, where knowledge has to be used rather than just learned.

Learners who can be grouped as reflective thinkers tend to make careful observations before assimilating information. They learn best through lectures, which are visual and auditory.

Kinaesthetic learning takes place by the learner or the doers who are actually carrying out a physical activity, rather than listening to a lecture or merely watching a demonstration. It is also referred to as tactile learning. Learners who have a predominantly kinesthetic learning style are thought to be fundamental discovery learners. They realize things through doing, as opposed to having thought first before initiating action. They may struggle to learn by reading or listening.

Auditory learning is associated with hearing words or numbers. They learn by listening to lectures and participating in discussions. The most common mode for information exchange is speech and is classified as auditory in the VARK model.

For auditory learners, it could be beneficial to verbally repeat information they have read and provided audiotapes or other sound devices relating to a subject.

K0243: Knowledge of Organizational Training and Education Policies, Processes, and Procedures

Organizational training represents a long-term investment in building highly skilled employees. Evaluation of such training programs needs to determine program effectiveness according to the value employees, businesses, and industry apply to train. Quality indicators, measures, moreover, performance standards for individual programs need to be identified from the overall workplace community, rather than from only educational content.

Organizational training policies are dependent on the role of employees and the nature of the organization and its primary functions. Organizations need to devise strategies according to what they want and expect from their employees.

Education policies should be based on the understanding of the teaching profession and the professional development of learners as a coherent component with several, interconnected perspectives, which include business goals, learning needs, support structures, targeted job structures, etc.

K0245: Knowledge of Principles and Processes
for Conducting Training and Education Needs Assessment

For organizations, training and education needs, assessment should be focused around skills related to the organization's core business functions. Such assessment takes into account issues relating to employee and organizational performance to establish whether training can help or not. Despite the crucial nature of needs assessment in training design and development, numerous studies have suggested that many training programs are inadequately planned, and in particular, are designed without the proper assessment of training needs. Managers should also understand that training needs assessment is a cycle process that should be re-evaluated and re-assessed periodically, Fig. 11.4 [10].

A needs assessment is a process that can be used to determine organizations' awareness and training needs. The results of the needs assessment can justify convincing management to allocate adequate resources to meet the identified awareness and training needs. In conducting a needs assessment, it is essential that critical relevant personnel be involved.

This first step (i.e., the need assessment) is then followed by a set of recommendations of topics to be covered, followed by preliminary course design and review. Review comments can then be used to finalize courses' content and design. Delivery of the curriculum should be followed by a feedback step to afford continuous improvement to the course materials and to identify additional educational needs.

After the implementation of a training program, metrics should be collected to monitor the accomplishment of the awareness and training program goals and objectives. This is accomplished by quantifying the level of implementation of awareness and training and the effectiveness and efficiency of the awareness and training. The evaluation process should also include analyzing the adequacy of awareness and training efforts and identifying possible improvements.

Fig. 11.4 The training cycle process [10]

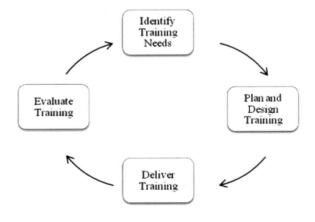

References

1. Bloom, B. S. (1956). *Taxonomy of educational objectives* (Vol. 1, pp. 20–24). Cognitive domain. New York: McKay.
2. Krathwohl, D. R., & Anderson, L. W. (2009). *A taxonomy for learning, teaching, and assessing: A revision of Bloom's taxonomy of educational objectives*. New York: Longman.
3. Anderson, L. W. (Ed.), Krathwohl, D. R. (Ed.), Airasian, P. W., Cruikshank, K. A., Mayer, R. E., Pintrich, P. R., Raths, J., & Wittrock, M. C. (2001). *A taxonomy for learning, teaching, and assessing: A revision of Bloom's taxonomy of educational objectives* (Complete ed.). New York: Longman.
4. LMS Data—Spring 2019 Updates. Retrieved March 17th, 2019, from https://edutechnica.com/2019/03/17/lms-data-spring-2019-updates/.
5. McIntosh, D. (2018). Vendors of Learning Management and eLearning Products. https://teachonline.ca/sites/default/files/pdfs/vendors_of_elearning_products_march2018.pdf.
6. Kolb, D. A. (1984). *Experiential learning: Experience as the source of learning and development*. Englewood Cliffs, NJ: Prentice-Hall.
7. McCarthy, B. (1987). The 4MAT system: Teaching to learning styles with right/left mode techniques. Excel, Incorporated.
8. Gardner, H. (1993). *Frames of mind: The theory of multiple intelligences*. New York, NY: Basic Books.
9. Barbe, W. B., & Milone, M. N. (February 1981). What we know about modality strengths. Educational Leadership. Association for Supervision and Curriculum Development, pp. 378–380.
10. Balderson, S. (2005). Strategy and human resource development. In J. P. Wilson & J. P. Wilson (Eds.), *Human resource development: Learning and training for individuals and organizations* (2nd ed., pp. 83–98). London: Kogan Page.

Chapter 12
Vulnerability Assessment and Management

Chuck Easttom

Introduction

Awareness of vulnerabilities is a critical issue in cybersecurity. There are well-known and documented vulnerabilities that can be readily addressed, but only if one is aware of them. Vulnerabilities are an issue of what you don't know can hurt you. Fortunately, there are a number of tools and resources that can aid you in understanding vulnerabilities in your system. You can then integrate that information into your security process.

The first step is to become aware of vulnerabilities in your network. As will be seen in this chapter, there is a wide array of tools and websites to assist you in this. But then that information must be integrated into your cybersecurity plans in order to mitigate the vulnerabilities. The latter part of this chapter will address that issue. This chapter is closely related to the material you saw in Chap. 10 and there is some minimal overlap.

Tools

In this section, we will examine some widely used tools for finding vulnerabilities. Some of these are general vulnerability scanners. Others check specific vulnerability issues. We will also look at both open source and commercial tools. The criteria for a tool being mentioned in this section is simply its popularity.

© The Editor(s) (if applicable) and The Author(s), under exclusive license to Springer Nature Switzerland AG 2020
I. Alsmadi et al., *The NICE Cyber Security Framework*,
https://doi.org/10.1007/978-3-030-41987-5_12

Shodan

Shodan is a well-known tool for vulnerability scanning. The website https://www.shodan.io/ is essentially a search engine for vulnerabilities. You need to sign up for a free account to use it, but there is no spam or other issued with the website. It is also popular with attackers, thus defenders should also use this site. You can also be sure that attackers use this site as well. You can see the website in Fig. 12.1.

The issue with Shodan is using the proper search terms. There are many you can use, a few will be discussed and one demonstrated here. Here are a few basic searches:

Search for default passwords with a specific filter

default password country:US
default password hostname:chuckeasttom.com
default password city:Dallas
default password state:AR

Find Apache servers

apache city:Dallas.

Find Webcams

webcamxp city:Houston
OLD IIS
"iis/6.0"

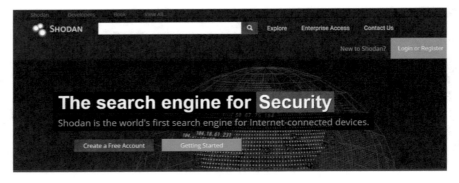

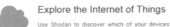

Fig. 12.1 Shodan.io

Fig. 12.2 Shodan search results

The preceding list are examples of search terms, the filters you can use include

- city: find devices in a specific city
- country: find devices in a specific country
- geo: you can pass it coordinates (i.e. latitude and longitude)
- hostname: find values that match a specific hostname
- net: search based on an IP or/x CIDR
- os: search based on operating system
- port: find particular ports that are open
- before/after: find results within a timeframe.

As an example, Fig. 12.2 shows the results for my search sslv3 state:tx

Of most use to cybersecurity professionals will be searching for vulnerabilities and filtering the results to a specific domain. That domain would be your organizations domain. Shodan is a very effective starting point for your vulnerability assessment. It will show you well-known vulnerabilities that are publicly accessible. This will be the first items you should address.

Maltego

Maltego is generally considered an open-source intelligence tool, rather than a vulnerability scanner. There are several versions of the product, some are free versions others are not. The website is https://www.paterva.com/web7/downloads.php#tab-3. The community version is free.

Results are well represented in a variety of easy to understand views In concert with its graphing libraries, Maltego identifies key relationships between data sets and identifies previously unknown relationships between them. Figure 12.3 shows the main screen of Maltego.

Maltego is often used for open-source intelligence. But it can be useful in understanding certain vulnerabilities. For example, if a suspect email is received, the website or email address can be searched with Maltego. Maltego is primarily used

Fig. 12.3 Maltego

by working with entities and transforms. You select some entity such as an email address or website and select a transform for that entity. Once you have selected something to graph, be it a person, email address, website or other items, the relationships between that entity and other entities are shown as a graph. This can be seen in Fig. 12.4.

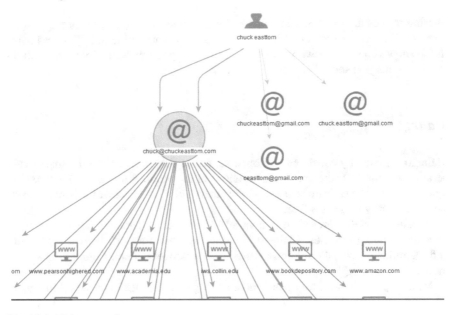

Fig. 12.4 Maltego graph

Maltego is more complex than some of the other tools we have discussed in this chapter. However, there are tutorials on the web to help you master this tool.

https://www.paterva.com/web7/docs/documentation.php
https://null-byte.wonderhowto.com/how-to/hack-like-pro-use-maltego-do-network-reconnaissance-0158464/

Spending some time learning the details of Maltego can be very worthwhile. It is not the most critical tool for vulnerability scanning but can be useful in analyzing vulnerabilities.

Nessus

Nessus (www.Nessus.org) is the most widely known commercial network vulnerability scanner. Tin the past there was a free version for personal use and a commercial version. It is now only available for a license cost. This is perhaps the most widely used vulnerability scanner available today. In this section, we will briefly explore the basic functionality. If you have an interest in learning more about Nessus, then it is recommended that you consult the documentation available at the Nessus website.

Nessus is a well-known vulnerability scanner. It has been used for many years. The license is currently over $2100 per year and can be obtained from https://www.tenable.com. Its price has been a barrier for small organizations with limited budgets. The primary advantage of Nessus is that the vendor is constantly updating the vulnerabilities it can scan for. Nessus also has a very easy-to-use web interface, as shown in Fig. 12.5.

If you select **New Scan**, you are given a number of options, as shown in Fig. 12.6.

You can select **Basic Network Scan** to see a number of intuitive basic settings. You have to name your scan and select a range of IP addresses, as shown in Fig. 12.7.

Fig. 12.5 Nessus new scan

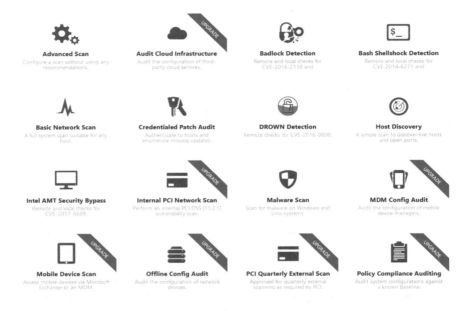

Fig. 12.6 Nessus options

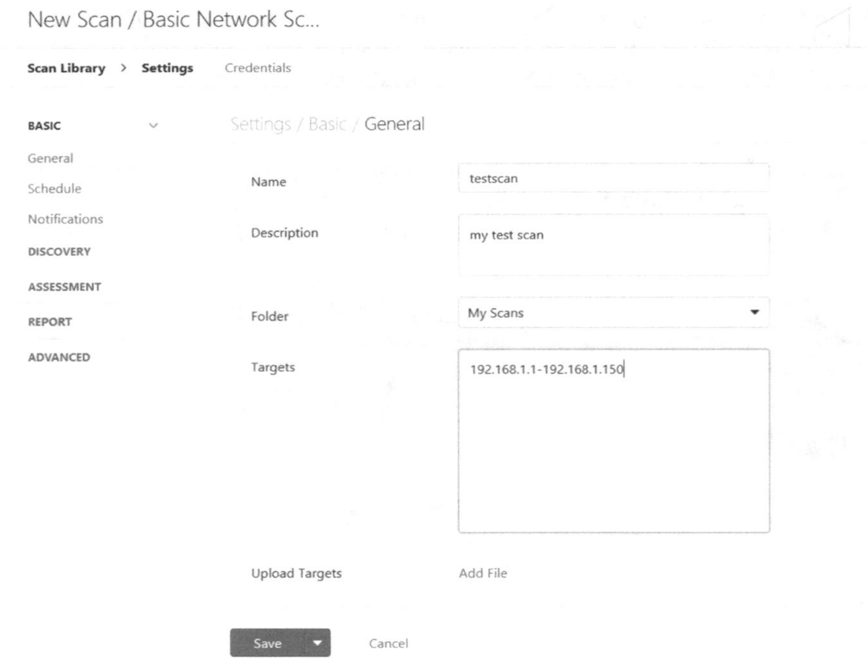

New Scan / Basic Network Sc...

Scan Library > **Settings**   Credentials

BASIC ⌄	Settings / Basic / General	
General		
Schedule	Name	testscan
Notifications		
DISCOVERY	Description	my test scan
ASSESSMENT		
REPORT	Folder	My Scans ▾
ADVANCED	Targets	192.168.1.1-192.168.1.150

Upload Targets   Add File

Save ▾   Cancel

Fig. 12.7 Nessus network scan

Then you can either schedule the scan to run later or launch it right away. Nessus scans can take some time to run because they are quite thorough. The results are presented in a very organized screen that is quite intuitive.

OWASP Zap

The Open Web Application Security Project (OWASP) is the standard for web application vulnerability. OWASP offers a free vulnerability scanner called the Zed Attack Proxy, commonly known as OWASP ZAP. You can download it from https://github. com/zaproxy/zaproxy/wiki/Downloads. The interface, shown in Fig. 12.8, is very easy to use.

The results are displayed in an easy to navigate format. One can simply double click on any specific result to get more details. The results can be seen in Fig. 12.9.

OWASP ZAP is a very easy-to-use tool. The basics can be mastered in a few minutes. And given that OWASP is the organization that tracks web application vulnerabilities, it is a very good source for testing the vulnerabilities of a website.

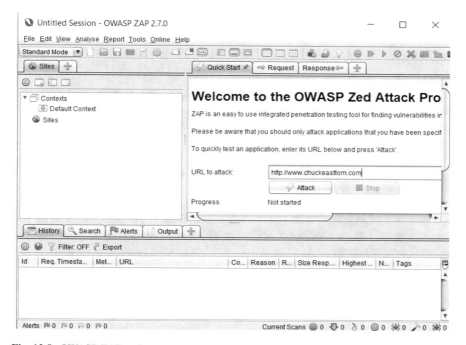

Fig. 12.8 OWASP ZAP main screen

Id	Req. Timesta...	Resp. Timest...	Met..	URL	C...	Reason	R...	Size Resp. H...	Size Resp....
700	10/22/19 1:28:...	10/22/19 1:28:...	GET	http://www.chuckeasttom.com/gr...	200	OK	2....	517 bytes	919,498 by...
701	10/22/19 1:28:...	10/22/19 1:28:...	GET	http://www.chuckeasttom.com/int...	200	OK	1....	517 bytes	322,519 by...
702	10/22/19 1:28:...	10/22/19 1:28:...	GET	http://www.chuckeasttom.com/gr...	200	OK	2....	517 bytes	919,498 by...
703	10/22/19 1:28:...	10/22/19 1:28:...	GET	http://www.chuckeasttom.com/int...	200	OK	2....	517 bytes	322,519 by...
704	10/22/19 1:28:...	10/22/19 1:28:...	GET	http://www.chuckeasttom.com/gr...	200	OK	2....	517 bytes	919,498 by...

Fig. 12.9 OWASP ZAP results

Fig. 12.10 OpenVAS online scan

OpenVAS

OpenVAS is one of the most widely used open-source vulnerability scanners. You can download OpenVAS, or you can use their online vulnerability scan. https://pentest-tools.com/network-vulnerability-scanning/network-security-scanner-online-openvas.

The website can be seen in Fig. 12.10.

The results are easy to see. Now on the result screen, you will be prompted to purchase the full version by upgrading to a pro account. However, you are also welcome to continue using the free version. This is shown in Fig. 12.11.

OpenVAS is an effective and easy-to-use tool. Even if you have other tools you are accustomed to, it would be advisable to include OpenVAS in your suite of vulnerability scanners.

Testing Specific Issues

In addition to general-purpose vulnerability scanners, there are websites that will scan for specific items. For example, the following site will provide you a report on the SSL/TLS security of any public website.

SSLLabs https://www.ssllabs.com/ssltest/analyze.html.

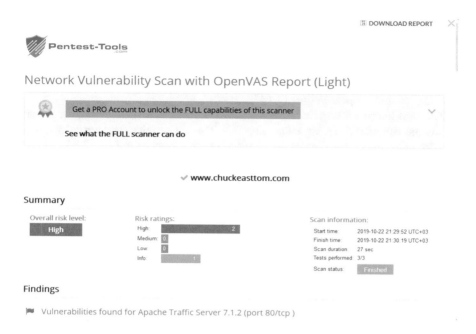

Fig. 12.11 OpenVAS results

You can see the results it provides in Fig. 12.12 .

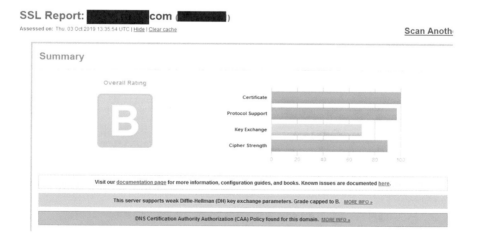

Fig. 12.12 SSL report

Another page that helps check SSL/TLS is Digicert. Digicert is also one of the major certificates authorities. You can see that page in Fig. 12.13. https://www. digicert.com/help/.

There are sites that also allow you to investigate specific threats. For example, Threatcrowd is a very popular thereat intelligence site https://www.threatcrowd.org. You can see that website in Fig. 12.14.

The Sans Internet Storm Center is another good resource for understanding current threats and vulnerabilities. https://isc.sans.edu. This site provides you a good resource for current events in threat research. You can see that website in Fig. 12.15.

✔ TLS Certificate

```
Common Name = www.myuhone.com
Issuer = COMODO RSA Organization Validation Secure Server CA
Serial Number = 10A4A3DEC44FCD60E0207FE4FED77DCA
SHA1 Thumbprint = 2070C66035185A98DC67B127685BE70E0DD752AC
Key Length = 2048
Signature algorithm = SHA256-RSA
Secure Renegotiation:
```

✔ TLS Certificate has not been revoked

OCSP Staple: Not Enabled

OCSP Origin: Good

CRL Status: Good

✔ TLS Certificate expiration

The certificate expires August 19, 2020 (321 days from today)

✔ Certificate Name matches www.myuhone.com

SERVER CERTIFICATE

Subject www.myuhone.com

Valid from 20/Aug/2019 to 19/Aug/2020

Issuer   COMODO RSA Organization Validation Secure Server CA

INTERMEDIATE CERTIFICATE

Subject COMODO RSA Organization Validation Secure Server CA

Valid from 12/Feb/2014 to 11/Feb/2029

Issuer   COMODO RSA Certification Authority

Fig. 12.13 DigiCert report

A Search Engine for Threats
SEARCH NOW >
Search by Domain, IP, Email or Organization

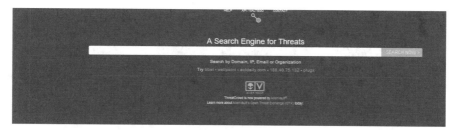

Fig. 12.14 Threatcrowd

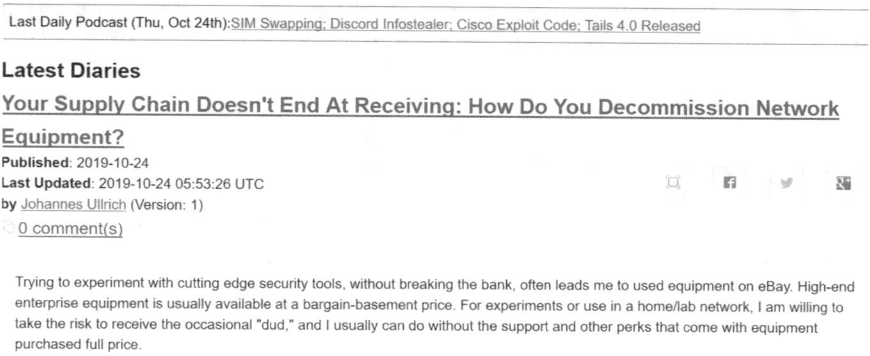

Last Daily Podcast (Thu, Oct 24th):SIM Swapping; Discord Infostealer; Cisco Exploit Code; Tails 4.0 Released
Latest Diaries
Your Supply Chain Doesn't End At Receiving: How Do You Decommission Network Equipment?
Published: 2019-10-24
Last Updated: 2019-10-24 05:53:26 UTC
by Johannes Ullrich (Version: 1)
0 comment(s)
Trying to experiment with cutting edge security tools, without breaking the bank, often leads me to used equipment on eBay. High-end enterprise equipment is usually available at a bargain-basement price. For experiments or use in a home/lab network, I am willing to take the risk to receive the occasional "dud," and I usually can do without the support and other perks that come with equipment purchased full price.

Fig. 12.15 Sans internet storm center

The Sans institute is well-known for publishing cybersecurity research. The storm center aggregates cybersecurity news. It could be argued that a cybersecurity professional should frequently reference the storm center (or some similar source) to keep current with ongoing threats.

Recon-Ng

This tool is an open-source Linux tool. It can be downloaded to any Linux machine. However, it comes with Kali Linux. You will see Kali later in this chapter in reference to Metasploit. This tool is a web reconnaissance tool written in Python. It has several different modules one can load and scan for specific vulnerabilities. There are literally scores of vulnerabilities that can be scanned for. You can see Recon-ng in Fig. 12.16.

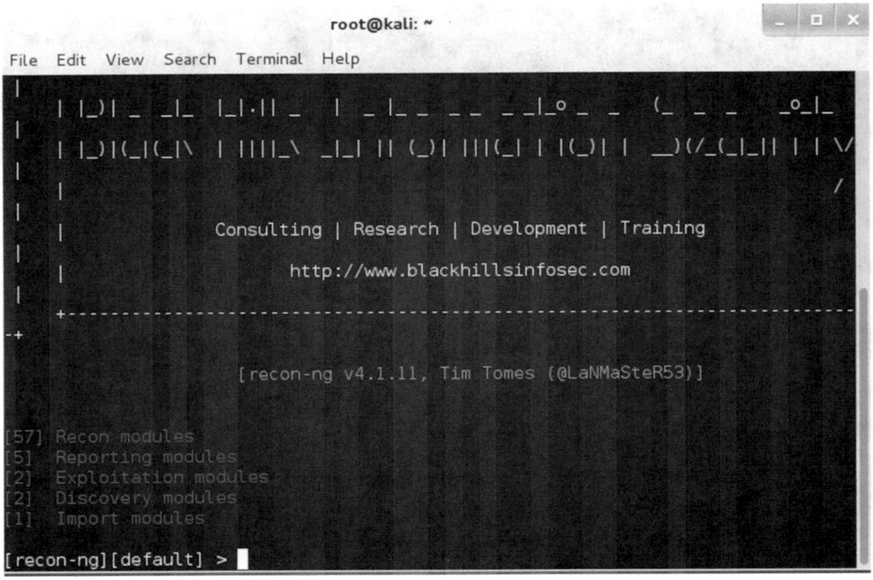

Fig. 12.16 Recon-ng

Metasploit

This tool is important enough to warrant a large separate section to itself. This tool can be downloaded separately, but ships with Kali Linux. Kali is a free Linux distribution. Metasploit is a tool often associated with penetration testing. It is most often described in the context of creating exploits to be delivered to a target. However, Metasploit has a large number of scanners built into it. These can be very useful in vulnerability assessment.

Much of Metasploit can be divided into four types of objects you will work with:

- Exploits: These are pieces of code that will attack a specific vulnerability.
- Payload: This is the code you actually send to the target. It is what actually does the dirty work on that target machine, once the exploit gets you in.
- Auxiliary: These modules provide some extra functionality. For example, scanning. For the purposes of this chapter, we will focus on these Auxiliary modules.
- Encoders: These embed exploits into other files like PDF, AVI, etc. We will see those in the next chapter.

It may be helpful to consider a quote from Rapid 7, the company that distributes Metasploit:

A vulnerability is a security hole in a piece of software, hardware or operating system that provides a potential angle to attack the system. A vulnerability can be as simple as weak passwords or as complex as buffer overflows or SQL injection vulnerabilities

"To take advantage of a vulnerability, you often need an exploit, a small and highly specialized computer program whose only reason of being is to take advantage of a specific vulnerability and to provide access to a computer system. Exploits often deliver a payload to the target system to grant the attacker access to the system.

The Metasploit Project host the world's largest public database of quality-assured exploits. Have a look at our exploit database – it's right here on the site" (https://community.rapid7.com/docs/DOC-2248)

Even if you don't use Metasploit as a tool for scanning vulnerabilities, visiting the website can keep you updated on existing security issues. In the coming sub-sections, some of the specific Metasploit scanners will be examined briefly.

SMB Scanner

Scanning for SMB is a very important SMB, Server Message Block, is used by Windows Active Directory. When you are scanning for this, you are checking to see if the target is a Windows computer that has SMB running. Now SMB is not a vulnerability per se. In fact, it is necessary for Windows Active Directory. However, there have been numerous exploits of flaws in SMB. The Wannacry virus exploited an SMB vulnerability. The scan is easy:

```
use scanner/smb/smb_version
set RHOSTS 192.168.1.177
set THREADS 4
run
```

Of course, you should replace the IP address 192.168.1.177 with the IP address of the target you are scanning. You can see the results in Fig. 12.17.

While this is simple, it has a lot of information in it. First is simply loading the specific scanner:

```
use scanner/smb/smb_version
```

```
msf auxiliary(smb_version) > set RHOSTS 192.168.1.177
RHOSTS => 192.168.1.177
msf auxiliary(smb_version) > set THREADS 4
THREADS => 4
msf auxiliary(smb_version) > run

[*] 192.168.1.177:445 is running Windows 2012 Standard Evaluation (build:9200)
name:WIN-7EP9LVQV307) (domain:WIN-7EP9LVQV307)
[*] Scanned 1 of 1 hosts (100% complete)
[*] Auxiliary module execution completed
msf auxiliary(smb_version) >
```

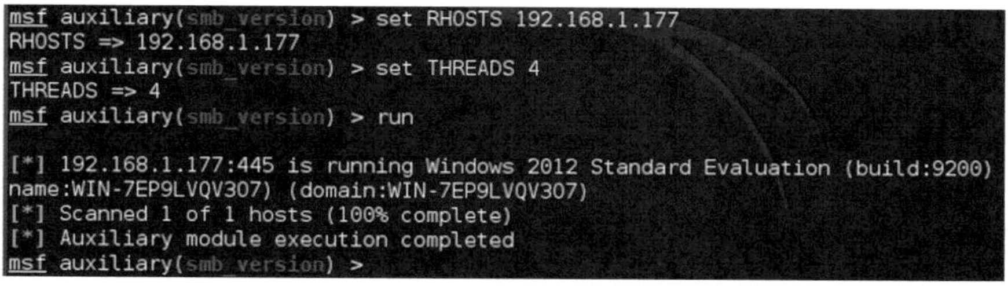

Fig. 12.17 SMB scanner

This is essentially how one loads any module in Metasploit. This is saying that you intend to use a specific module. And it gives the path to that module. Notice the first part of the path is *scanner*. This particular directory has a number of scanners you can use. The next line is:

set RHOSTS 192.168.1.177

First, notice the RHOSTS. This is the IP address for the remote host(s) you are scanning. Some modules will have RHOST, indicating you can only scan one target, others will have RHOSTS, indicating you can scan several targets if you wish. All scanners will use an RHOST or RHOSTS, but not all exploit modules will Anytime a module has RHOSTS rather than RHOST, you could scan a range of IP addresses. Just modify the command to say:

set RHOSTS 192.168.1.177 192.168.1.215

Then we have

set THREADS 4

This is telling Metasploit how many threads to use to run this module. There is no specific rule on this, other than don't select too high a number or your own machines CPU may not be able to handle it. When in doubt, just go with 1 thread. Finally, we have:

run

Every module om Metasploit ends with either *run* or *exploit*. This just tells Metasploit to go and do whatever you have just setup. If the target does not have the vulnerability you are scanning for, then you will get no results.

SQL Server Scan

This is another scanner that is not inherently a vulnerability. Having SQL Server is an important part of many Microsoft networks. However, being aware that the machines are running and may require patching can be useful. If you carefully studied the SMB scan, then the commands here will be obvious. You type in:

use auxiliary/scanner/mssql/mssql_ping
set RHOSTS 192.168.1.177
Set THREADS 1
Set USE_WINDOWS_AUTHENT false
run

There is only one new item, which is USE_WINDOWS_AUTHENT false. This is just telling Metasploit that you don't have any login credentials for SQL Server, so don't attempt to login The results can be seen in Fig. 12.18.

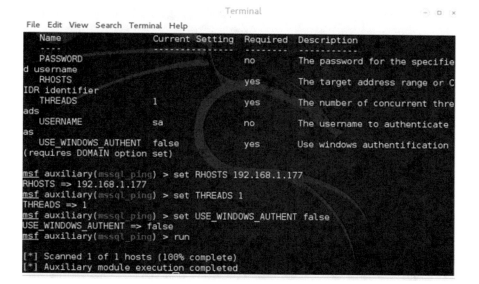

Fig. 12.18 SQL server scan

This is a very simple scanner to run. As we examine other scanners, you will probably notice some commonalities in all the various scanners.

SSH Server Scan

This is a scan to detect SSH (Secure shell) servers on the target. SSH is a secure remote access protocol. So, you may wonder why it would be scanned for. Certainly, you want administrators to utilize SSH rather than options such as telnet. The issue is unauthorized SSH communication. Since SSH is encrypted, it is also an excellent way for a malicious insider to exfiltrate data. That makes it important that you know of any SSH processes on your network. The commands are very similar to what you have already seen:

```
use scanner/ssh/ssh_version
set RHOSTS 192.168.1.177
Set THREADS 1
Set USE_WINDOWS_AUTHENT false
Run
```

Since all of these commands have already been explained, no further explanation is needed. You can see the results in Fig. 12.19.

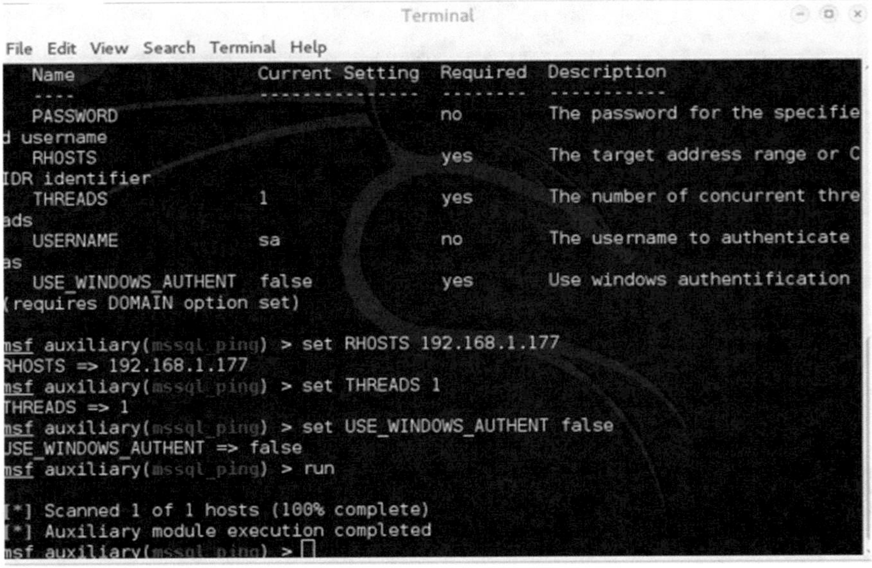

```
                                          Terminal                           - □ ×
File  Edit  View  Search  Terminal  Help
   Name                    Current Setting   Required  Description
   ----                    ---------------   --------  -----------
   PASSWORD                                  no        The password for the specifie
d username
   RHOSTS                                    yes       The target address range or C
IDR identifier
   THREADS                 1                 yes       The number of concurrent thre
ads
   USERNAME                sa                no        The username to authenticate
as
   USE_WINDOWS_AUTHENT     false             yes       Use windows authentification
(requires DOMAIN option set)

msf auxiliary(mssql_ping) > set RHOSTS 192.168.1.177
RHOSTS => 192.168.1.177
msf auxiliary(mssql_ping) > set THREADS 1
THREADS => 1
msf auxiliary(mssql_ping) > set USE_WINDOWS_AUTHENT false
USE_WINDOWS_AUTHENT => false
msf auxiliary(mssql_ping) > run

[*] Scanned 1 of 1 hosts (100% complete)
[*] Auxiliary module execution completed
msf auxiliary(mssql_ping) > □
```

Fig. 12.19 SSH scan

Anonymous FTP Servers

As you might guess, this scans for FTP servers that allow anonymous login. This is a significant vulnerability. Many FTP servers use anonymous logins by default. You need to be aware of when such servers are running on your network,

> use auxiliary/scanner/ftp/anonymous
> set RHOSTS 192.168.1.177
> Set RPORT 21
> Set THREADS 1
> Set USE_WINDOWS_AUTHENT false
> run

You might notice that some scans are in the/scanner/directory, but many are in the/auxiliary/scanner/directory. You can see the results of this scan in Fig. 12.20.

Other Tools

Metasploit also integrates with other tools. For example, nmap is integrated into Metasploit. It is also possible to integrate OpenVAS into Metasploit. These can both be executed without Metasploit. However, integrating these tools along with

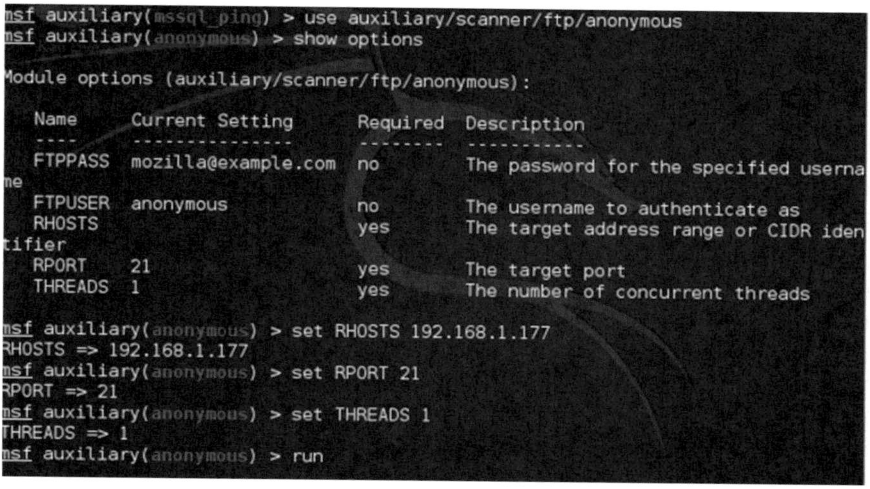

```
msf auxiliary(mssql_ping) > use auxiliary/scanner/ftp/anonymous
msf auxiliary(anonymous) > show options

Module options (auxiliary/scanner/ftp/anonymous):

   Name       Current Setting      Required  Description
   ----       ---------------      --------  -----------
   FTPPASS    mozilla@example.com  no        The password for the specified userna
me
   FTPUSER    anonymous            no        The username to authenticate as
   RHOSTS                          yes       The target address range or CIDR iden
tifier
   RPORT      21                   yes       The target port
   THREADS    1                    yes       The number of concurrent threads

msf auxiliary(anonymous) > set RHOSTS 192.168.1.177
RHOSTS => 192.168.1.177
msf auxiliary(anonymous) > set RPORT 21
RPORT => 21
msf auxiliary(anonymous) > set THREADS 1
THREADS => 1
msf auxiliary(anonymous) > run
```

Fig. 12.20 Anonymous FTP scanner

Metasploit gives a single location to run scans from. Also, one can setup Metasploit to record all scans into a database. This allows for reporting of all these scans by querying that database.

Responding to Vulnerability

Now that you have a range of tools to find vulnerabilities, the issue becomes how to respond. This will depend on the nature of the vulnerability. In the case that the issue is simply a missing patch, such as an outdated version of SMB, then the obvious response is to first test the patch, then roll it out to your entire network.

Some issues, however, require a great deal of resources to remediate. It may seem odd, but then one has to evaluate whether not to remediate the issue. There is a basic formula one works though. One quantifies the severity of the issue and the probability of it occurring. The combination of severity and probability gives you an indication of how substantial a risk the vulnerability presents. Chap. 10 discussed CVSS in some detail. Then you compare that with the resources required to address the issue. If the resources required to address an issue then there has to be a decision made as to whether or not to expend the resources.

There are always four possible responses to any issue:

Mitigation: Take steps to reduce the impact or reduce the probability of the event occurring. Anti-virus software is a classic example. It reduces the risk of a virus infection occurring.
Avoidance: This is eliminating the risk. Often this is simply not possible.

Transfer: Transferring the risk means that someone else is now responsible should the event occur. This is often seen as cyber breach insurance. Normally, the insurance carrier will require certain mitigating controls to be in place.

Acceptance: This is very risky. This is essentially stating that addressing the issue costs more than the impact of the risk is realized, or that the probability of it is so remote that it can be ignored. Acceptance should only be embraced after careful threats.

The most common response to any vulnerability or risk is mitigation. How much resources are allocated to mitigation is contingent upon the likelihood of the risk being realized and the impact should it be realized.

Risk and vulnerability response begins by ranking each risk based on probability and impact. Then those vulnerabilities are ordered based on their ranking. The highest-ranking vulnerabilities should be addressed first. Addressing vulnerabilities is much like any other process in cybersecurity. It should be carefully planned and thought out. It is not an ad hoc addressing vulnerability with no order or plan.

Conclusions

In this chapter, you have seen a wide range of tools and techniques for vulnerability scanning. It is important that all security professionals be very aware of the vulnerabilities in their systems. Each of the tools described in this chapter can be part of your vulnerability scanning approach. We also covered briefly how to respond once you find a vulnerability.

Index